HEATH

ALGEBRA 1

AN INTEGRATED APPROACH

LARSON, KANOLD, STIFF

LESSON PLANS

McDougal Littell

Evanston, Illinois • Boston • Dallas

International Standard Book Number: 0-395-87938-8

1 2 3 4 5 6 7 8 9 10 HWI 01 00 99 98 97

Lesson Plan 1.1

pages 2–7

1 day for Basic Course, 1 day for Full Course

Teacher's Name ____________________ *Class* __________ *Date* __________ *Room* __________

Goals
1. Represent numbers and number operations.
2. Use grouping symbols.

State/Local Objectives ____________________

NCTM Curriculum Standards: Problem Solving, Communication, Reasoning, Connections

✓ **Check items you wish to use for this lesson.**

Introducing the Lesson
- ___ Teaching Tools: Problem of the Day copymasters page 21, or Teacher's Edition page 2
- ___ Teaching Tools: Warm-Up Exercises copymasters page 52, or Teacher's Edition page 2

Teaching the Lesson using the following:
- ___ Problem Solving Activity for Example 2, Teacher's Edition page 3
- ___ Alternative Approach for Example 3, Teacher's Edition page 4
- ___ Color Transparencies pages 1, 2
- ___ Teaching Tools: transparencies page 5

Notes for substitute teacher ____________________

Closing the Lesson
- ___ Communicating about Algebra, Student's Edition page 4
- ___ Communicating about Algebra, Teacher's Edition page 4
- ___ Guided Practice Exercises, Student's Edition page 5

Homework Assignment, pages 5–7
- ___ Basic/Average: Ex. 7–14, 15–33 odd, 35, 39 odd, 41–42, 43–53 odd, 55–58
- ___ Above Average: Ex. 7–14, 19–33 odd, 35–39 odd, 41–42, 43–53 odd, 55–58
- ___ Advanced: Ex. 7–13 odd, 19–33 odd, 35–42, 51–58

Reteaching the Lesson
- ___ Extra Practice Copymasters page 1
- ___ Reteaching Copymasters page 1
- ___ Alternative Assessment, Math Log pages 27–30

Extending the Lesson
- ___ Writing, Teacher's Edition page 7
- ___ Problem Solving Activity for Exercises 55–58, Teacher's Edition page 7
- ___ Technology: Using Calculators and Computers pages 1–6

Notes

Lesson Plan 1.2
pages 8–13

1 day for Basic Course, 1 day for Full Course

Teacher's Name ______________ *Class* ______________ *Date* ______________ *Room* ______________

Goals 1. Evaluate variable expressions.
2. Represent real-life quantities using variable expressions.

State/Local Objectives ______________

NCTM Curriculum Standards: Problem Solving, Communication, Reasoning, Connections, Geometry, Technology

✓ **Check items you wish to use for this lesson.**

Introducing the Lesson
___ Teaching Tools: Problem of the Day copymasters page 21, or Teacher's Edition page 8
___ Teaching Tools: Warm-Up Exercises copymasters page 52, or Teacher's Edition page 8

Notes for substitute teacher ______________

Closing the Lesson
___ Communicating about Algebra, Student's Edition page 10
___ Cooperative Learning: Communicating about Algebra, Teacher's Edition page 10
___ Extension: Communicating about Algebra, Teacher's Edition page 10
___ Guided Practice Exercises, Student's Edition page 11

Homework Assignment, pages 11–13
___ Basic/Average: Ex. 7–15 odd, 17–24, 25–45 odd, 47–50
___ Above Average: Ex. 7–41 odd, 43–50
___ Advanced: Ex. 7–23 odd, 25–28, 43–50

Reteaching the Lesson
___ Extra Practice Copymasters page 2
___ Reteaching Copymasters page 2

Extending the Lesson
___ Group Activity, Teacher's Edition page 12
___ Enrichment, Teacher's Edition page 12
___ Technology: Using Calculators and Computers page 7

Notes

Lesson Plan 1.3
pages 14–18

1 day for Basic Course, 1 day for Full Course

Teacher's Name ______________________ *Class* __________ *Date* __________ *Room* __________

Goals
1. Evaluate expressions containing exponents.
2. Use exponents.

State/Local Objectives ______________________________

NCTM Curriculum Standards: Problem Solving, Communication, Geometry, Connections, Technology

✓ **Check items you wish to use for this lesson.**

Introducing the Lesson
___ Teaching Tools: Problem of the Day copymasters page 21, or Teacher's Edition page 14
___ Teaching Tools: Warm-Up Exercises copymasters page 52, or Teacher's Edition page 14

Teaching the Lesson using the following:
___ Common-Error Alert, Teacher's Edition page 14
___ Teaching Tools: transparencies pages 1, 2

Notes for substitute teacher ______________________________

Closing the Lesson
___ Communicating about Algebra, Student's Edition page 16
___ Guided Practice Exercises, Student's Edition page 17

Homework Assignment, pages 17–18
___ Basic/Average: Ex. 7–15, 17–45 odd
___ Above Average: Ex. 7–23 odd, 24, 25–35 odd, 41, 43–46
___ Advanced: Ex. 7–27 odd, 32–36, 41–46

Reteaching the Lesson
___ Extra Practice Copymasters page 3
___ Reteaching Copymasters page 3

Extending the Lesson
___ Enrichment, Teacher's Edition page 18
___ Research, Teacher's Edition page 18

Notes

Lesson Plan 1.4
pages 20–25

1 day for Basic Course, 1 day for Full Course

Teacher's Name ______________________ *Class* __________ *Date* __________ *Room* __________

Goals
1. Use order of operations to evaluate algebraic expressions.
2. Evaluate expressions with a calculator.

State/Local Objectives ____

NCTM Curriculum Standards: Problem Solving, Communication, Reasoning, Connections, Geometry, Technology

✓ **Check items you wish to use for this lesson.**

Introducing the Lesson
___ Teaching Tools: Problem of the Day copymasters page 3, or Teacher's Edition page 20
___ Teaching Tools: Warm-Up Exercises copymasters page 52, or Teacher's Edition page 20

Teaching the Lesson using the following:
___ Common-Error Alert, Teacher's Edition page 22
___ Color Transparencies page 2

Notes for substitute teacher ____

Closing the Lesson
___ Communicating about Algebra, Student's Edition page 22
___ Problem Solving: Communicating about Algebra, Teacher's Edition page 22
___ Guided Practice Exercises, Student's Edition page 23

Homework Assignment, pages 23–25
___ Basic/Average: Ex. 5–39 odd, 41–45 odd
___ Above Average: Ex. 17–39 odd, 40–45
___ Advanced: Ex. 17–29 odd, 31–35, 39–45

Reteaching the Lesson
___ Extra Practice Copymasters page 4
___ Reteaching Copymasters page 4

Extending the Lesson
___ Enrichment, Teacher's Edition page 25

Notes

Lesson Plan 1.5
pages 27–32

1 day for Basic Course, 1 day for Full Course

Teacher's Name ______________________ *Class* __________ *Date* __________ *Room* ________

Goals
1. Check and solve equations.
2. Check solutions of inequalities.

State/Local Objectives ____

NCTM Curriculum Standards: Problem Solving, Communication, Reasoning, Connections

✓ **Check items you wish to use for this lesson.**

Introducing the Lesson
___ Teaching Tools: Problem of the Day copymasters page 22, or Teacher's Edition page 27
___ Teaching Tools: Warm-Up Exercises copymasters page 52, or Teacher's Edition page 27

Teaching the Lesson using the following:
___ Math Journal, Teacher's Edition page 28
___ Writing, Teacher's Edition page 29

Notes for substitute teacher ____

Closing the Lesson
___ Communicating about Algebra, Student's Edition page 29
___ Cooperative Learning: Communicating about Algebra, Teacher's Edition page 29
___ Guided Practice Exercises, Student's Edition page 3

Homework Assignment, pages 30–32
___ Basic/Average: Ex. 9–30 multiples of 3, 32, 33–57 odd, 60–63
___ Above Average: Ex. 9–30 multiples of 3, 32, 33–57 odd, 59–65 odd
___ Advanced: Ex. 9–30 multiples of 3, 32, 33–36, 51–65 odd, 66

Reteaching the Lesson
___ Extra Practice Copymasters page 5
___ Reteaching Copymasters page 5

Extending the Lesson
___ Enrichment, Teacher's Edition page 31
___ Group Activity, Teacher's Edition page 31
___ Group Activity, Teacher's Edition page 32

Notes

Lesson Plan 1.6
pages 33–38

2 days for Basic Course, 1 day for Full Course

Teacher's Name ________________ *Class* ________ *Date* ________ *Room* ________

Goals
1. Translate verbal phrases into algebraic expressions.
2. Translate verbal sentences into algebraic equations and inequalities.

State/Local Objectives ____

____

NCTM Curriculum Standards: Problem Solving, Communication, Reasoning, Connections, Geometry

✓ **Check items you wish to use for this lesson.**

Introducing the Lesson
___ Teaching Tools: Problem of the Day copymasters page 22, or Teacher's Edition page 33
___ Teaching Tools: Warm-Up Exercises copymasters page 53, or Teacher's Edition page 33

Teaching the Lesson using the following:
___ Math Journal, Teacher's Edition page 35

Notes for substitute teacher ____

____

____

Closing the Lesson
___ Communicating about Algebra, Student's Edition page 35
___ Guided Practice Exercises, Student's Edition page 36

Homework Assignment, pages 36–38
___ Basic/Average: Ex. 9–33 odd, 39–53 odd
___ Above Average: Ex. 9–33 odd, 39–53 odd
___ Advanced: Ex. 9–25 odd, 26, 31–34, 39–55 odd, 56

Reteaching the Lesson
___ Extra Practice Copymasters page 6
___ Reteaching Copymasters page 6

Extending the Lesson
___ Enrichment, Teacher's Edition page 37
___ Cooperative Learning, Teacher's Edition page 38
___ Applications Handbook pages 40, 56

Notes

____

____

____

Lesson Plan 1.7
pages 39–45

2 days for Basic Course, 1 day for Full Course

*Teacher's Name*____________________ *Class*________ *Date*________ *Room*________

Goals
1. Use algebra to solve simple real-life problems.
2. Make an algebraic model for a real-life problem.

State/Local Objectives ____

____

NCTM Curriculum Standards: Problem Solving, Communication, Reasoning, Connections

✓ **Check items you wish to use for this lesson.**

Introducing the Lesson
___ Teaching Tools: Problem of the Day copymasters page 23, or Teacher's Edition page 39
___ Teaching Tools: Warm-Up Exercises copymasters page 53, or Teacher's Edition page 39

Teaching the Lesson using the following:
___ Math Journal, Teacher's Edition page 41
___ Teaching Tools: transparencies page 3

Notes for substitute teacher ____

____

____

Closing the Lesson
___ Communicating about Algebra, Student's Edition page 41
___ Problem Solving: Communicating about Algebra, Teacher's Edition page 42
___ Guided Practice Exercises, Student's Edition page 42

Homework Assignment, pages 42–45
___ Basic/Average: Ex. 7–24, 25–35
___ Above Average: Ex. 7–24, 25–36
___ Advanced: Ex. 7–28, 29–42

Reteaching the Lesson
___ Extra Practice Copymasters page 7
___ Reteaching Copymasters page 7

Extending the Lesson
___ Enrichment, Teacher's Edition page 45
___ Cultural Diversity Extensions page 1

Notes

____

____

____

Lesson Plan 1.8
pages 49–54

1 day for Basic Course, 1 day for Full Course

Teacher's Name ______________________ *Class* __________ *Date* __________ *Room* __________

Goals
1. Use tables to organize data.
2. Use graphs to model data visually.

State/Local Objectives ____

NCTM Curriculum Standards: Problem Solving, Communication, Reasoning, Connections, Discrete Mathematics, Statistic

✓ **Check items you wish to use for this lesson.**

Introducing the Lesson
___ Teaching Tools: Problem of the Day copymasters page 23, or Teacher's Edition page 49
___ Teaching Tools: Warm-Up Exercises copymasters page 53, or Teacher's Edition page 49

Teaching the Lesson using the following:
___ Common-Error Alert, Teacher's Edition page 50
___ Math Journal, Teacher's Edition page 51
___ Color Transparencies pages 3, 4, 5, 6
___ Teaching Tools: transparencies page 4

Notes for substitute teacher ____

Closing the Lesson
___ Communicating about Algebra, Student's Edition page 51
___ Guided Practice Exercises, Student's Edition page 52

Homework Assignment, pages 52–54
___ Basic/Average: Ex. 5–20
___ Above Average: Ex. 5–23
___ Advanced: Ex. 5–25

Reteaching the Lesson
___ Extra Practice Copymasters page 8
___ Reteaching Copymasters page 8

Extending the Lesson
___ Career Connections, Teacher's Edition page 53
___ Enrichment, Teacher's Edition page 54
___ Applications Handbook pages 40, 53
___ Technology: Using Calculators and Computers page 9

Notes

Lesson Plan 2.1
pages 62–67

1 day for Basic Course, 1 day for Full Course

Teacher's Name ______________ *Class* ______ *Date* ______ *Room* ______

Goals

1. Graph and compare real numbers.
2. Find opposites and absolute values.

State/Local Objectives ______________

NCTM Curriculum Standards: Problem Solving, Communication, Reasoning, Connections, Geometry, Structure

✓ **Check items you wish to use for this lesson.**

Introducing the Lesson
___ Teaching Tools: Problem of the Day copymasters page 23, or Teacher's Edition page 62
___ Teaching Tools: Warm-Up Exercises copymasters page 54, or Teacher's Edition page 62

Teaching the Lesson using the following:
___ Alternative Approach for Example 2, Teacher's Edition page 63
___ Common-Error Alert, Teacher's Edition page 64
___ Color Transparencies page 8
___ Teaching Tools: transparencies page 6

Notes for substitute teacher ______________

Closing the Lesson
___ Communicating about Algebra, Student's Edition page 65
___ Guided Practice Exercises, Student's Edition page 66

Homework Assignment, pages 66–67
___ Basic/Average: Ex. 7–27 odd, 31–32, 35–36, 39–51 odd
___ Above Average: Ex. 9–27 multiples of 3, 29–32, 39–53 odd
___ Advanced: Ex. 29–32, 37–38, 39–47 odd, 49–56

Reteaching the Lesson
___ Extra Practice Copymasters page 9
___ Reteaching Copymasters page 9
___ Alternative Assessment: Math Log pages 31–33

Extending the Lesson
___ Writing, Teacher's Edition page 67
___ Technology: Using Calculators and Computers page 11

Notes

Lesson Plan 2.2
pages 68–73

1 day for Basic Course, 2 days for Full Course

Teacher's Name ______________________ *Class* ____________ *Date* __________ *Room* ________

Goals
1. Add real numbers.
2. Add real numbers using a calculator.

State/Local Objectives ____

NCTM Curriculum Standards: Problem Solving, Communication, Reasoning, Connections, Structure, Geometry, Technology

✓ **Check items you wish to use for this lesson.**

Introducing the Lesson
___ Teaching Tools: Problem of the Day copymasters page 24, or Teacher's Edition page 68
___ Teaching Tools: Warm-Up Exercises copymasters page 54, or Teacher's Edition page 68

Teaching the Lesson using the following:
___ Alternative Approach for Example 2, Teacher's Edition page 68
___ Alternative Approach for Example 3, Teacher's Edition page 69
___ Technology for Example 5, Teacher's Edition page 69
___ Color Transparencies pages 9, 10
___ Teaching Tools: transparencies page 6

Notes for substitute teacher ____

Closing the Lesson
___ Communicating about Algebra, Student's Edition page 70
___ Extend Communicating: Cooperative Learning, Teacher's Edition page 71
___ Guided Practice Exercises, Student's Edition page 71

Homework Assignment, pages 71–73
___ Basic/Average: Ex. 5–31 odd, 33–42 multiples of 3, 45–51 odd, 54–56
___ Above Average: Ex. 6–30 multiples of 3, 33–39 odd, 45–55 odd
___ Advanced: Ex. 6–30 multiples of 3, 33–43 odd, 51–53, 57

Reteaching the Lesson
___ Extra Practice Copymasters page 10
___ Reteaching Copymasters page 10

Extending the Lesson
___ Historical Note, Teacher's Edition page 69
___ Writing, Teacher's Edition page 71
___ Enrichment, Teacher's Edition page 73
___ Technology: Using Calculators and Computers page 13

Notes

Lesson Plan 2.3
pages 75–80

1 day for Basic Course, 1 day for Full Course

Teacher's Name ____________________ *Class* __________ *Date* __________ *Room* __________

Goals
1. Subtract real numbers.
2. Simplify the difference of two algebraic expressions.

State/Local Objectives ____________________

NCTM Curriculum Standards: Problem Solving, Communication, Reasoning, Connections, Structure

✓ **Check items you wish to use for this lesson.**

Introducing the Lesson
___ Teaching Tools: Problem of the Day copymasters page 24, or Teacher's Edition page 75
___ Teaching Tools: Warm-Up Exercises copymasters page 54, or Teacher's Edition page 75

Teaching the Lesson using the following:
___ Alternative Approach for Example 1, Teacher's Edition pages 75–76
___ Common-Error Alert, Teacher's Edition page 76

Notes for substitute teacher ____________________

Closing the Lesson
___ Communicating about Algebra, Student's Edition page 77
___ Extend Communicating: Cooperative Learning, Teacher's Edition page 78
___ Guided Practice Exercises, Student's Edition page 78

Homework Assignment, pages 78–79
___ Basic/Average: Ex. 9–42 multiples of 3, 43, 45–46, 48–57 multiples of 3, 59, 62
___ Above Average: Ex. 9–42 multiples of 3, 45–50, 59–64
___ Advanced: Ex. 24–42 multiples of 3, 44–50, 61–66

Reteaching the Lesson
___ Extra Practice Copymasters page 11
___ Reteaching Copymasters page 11

Extending the Lesson
___ Applications Handbook pages 58–59

Notes

Lesson Plan 2.4
pages 81–86

1 day for Basic Course, 1 day for Full Course

Teacher's Name ______________________ *Class* __________ *Date* __________ *Room* __________

Goals
1. Organize data in a matrix.
2. Add and subtract matrices.

State/Local Objectives ______________________

NCTM Curriculum Standards: Problem Solving, Communication, Connections, Discrete Mathematics, Statistics

✓ **Check items you wish to use for this lesson.**

Introducing the Lesson
___ Teaching Tools: Problem of the Day copymasters page 24, or Teacher's Edition page 81
___ Teaching Tools: Warm-Up Exercises copymasters page 55, or Teacher's Edition page 81

Teaching the Lesson using the following:
___ Teaching Tools: transparencies pages 1, 2

Notes for substitute teacher ______________________

Closing the Lesson
___ Communicating about Algebra, Student's Edition page 83
___ Extend Communicating: Cooperative Learning, Teacher's Edition page 83
___ Guided Practice Exercises, Student's Edition page 84

Homework Assignment, pages 84–86
___ Basic/Average: Ex. 5–19, 21–23 odd, 27–36
___ Above Average: Ex. 5–19, 21–24, 27–36
___ Advanced: Ex. 5–8, 9–23 odd, 25–26, 29–36

Reteaching the Lesson
___ Extra Practice Copymasters page 12
___ Reteaching Copymasters page 12

Extending the Lesson
___ Technology: Using Calculators and Computers page 16

Notes

Lesson Plan 2.5
pages 89–94

2 days for Basic Course, 1 day for Full Course

Teacher's Name ____________ *Class* ______ *Date* ______ *Room* ______

Goals
1. Multiply real numbers.
2. Use calculators and units of measure in products.

State/Local Objectives ____________

NCTM Curriculum Standards: Problem Solving, Communication, Reasoning, Connections, Structure, Geometry

✓ **Check items you wish to use for this lesson.**

Introducing the Lesson
___ Teaching Tools: Problem of the Day copymasters page 25, or Teacher's Edition page 89
___ Teaching Tools: Warm-Up Exercises copymasters page 55, or Teacher's Edition page 89
___ Technology for Example 4, Teacher's Edition page 90

Notes for substitute teacher ____________

Closing the Lesson
___ Communicating about Algebra, Student's Edition page 91
___ Extend Communicating: Cooperative Learning, Teacher's Edition page 92
___ Guided Practice Exercises, Student's Edition page 92

Homework Assignment, pages 92–94
___ Basic/Average: Ex. 7–27 odd, 31–34, 35–49 odd, 51–60, 61–65 odd
___ Above Average: Ex. 8–26 even, 29–34, 36–50 even, 59–60, 62–66
___ Advanced: Ex. 15–49 odd, 59–68

Reteaching the Lesson
___ Extra Practice Copymasters page 13
___ Reteaching Copymasters page 13

Extending the Lesson
___ Applications Handbook pages 60–61

Notes

Lesson Plan 2.6
pages 95–100

1 day for Basic Course, 1 day for Full Course

Teacher's Name ____________ *Class* ________ *Date* ________ *Room* ______

Goals
1. Use the Distributive Property.
2. Simplify expressions by combining like terms.

State/Local Objectives ____________

NCTM Curriculum Standards: Problem Solving, Communication, Reasoning, Connections, Structure, Geometry

✓ **Check items you wish to use for this lesson.**

Introducing the Lesson
___ Teaching Tools: Problem of the Day copymasters page 25, or Teacher's Edition page 95
___ Teaching Tools: Warm-Up Exercises copymasters page 55, or Teacher's Edition page 95

Teaching the Lesson using the following:
___ Common-Error Alert, Teacher's Edition page 96

Notes for substitute teacher ____________

Closing the Lesson
___ Communicating about Algebra, Student's Edition page 97
___ Extend Communicating: Cooperative Learning, Teacher's Edition page 97
___ Guided Practice Exercises, Student's Edition page 98

Homework Assignment, pages 98–100
___ Basic/Average: Ex. 9–36 multiples of 3, 37–45 odd, 47–48, 51–65 odd, 67–69
___ Above Average: Ex. 9–36 multiples of 3, 37–45 odd, 49–50, 51–54, 55–65 odd, 67–69
___ Advanced: Ex. 25–36, 38–46 even, 47–54, 59–64, 67–69

Reteaching the Lesson
___ Extra Practice Copymasters page 14
___ Reteaching Copymasters page 14

Extending the Lesson
___ Writing, Teacher's Edition page 97
___ Teaching Tools: transparencies pages 18, 19

Notes

Lesson Plan 2.7

pages 101–106

1 day for Basic Course, 1 day for Full Course

Teacher's Name ______________________ *Class* ________ *Date* ________ *Room* ________

Goals
1. Express division as multiplication.
2. Divide real numbers.

State/Local Objectives ______________________

NCTM Curriculum Standards: Problem Solving, Communication, Reasoning, Connections, Structure, Technology

✓ **Check items you wish to use for this lesson.**

Introducing the Lesson
___ Teaching Tools: Problem of the Day copymasters page 25, or Teacher's Edition page 101
___ Teaching Tools: Warm-Up Exercises copymasters page 56, or Teacher's Edition page 101

Teaching the Lesson using the following:
___ Common-Error Alert, Teacher's Edition page 103
___ Color Transparencies page 11

Notes for substitute teacher ______________________

Closing the Lesson
___ Communicating about Algebra, Student's Edition page 103
___ Extend Communicating: Cooperative Learning, Teacher's Edition page 103
___ Guided Practice Exercises, Student's Edition page 104

Homework Assignment, pages 104–106
___ Basic/Average: Ex. 9–36 multiples of 3, 37–40, 41–47 odd, 49–51, 57–73 odd, 75–78
___ Above Average: Ex. 9–36 multiples of 3, 38–48 even, 50, 53–56, 57–72 multiples of 3, 75–81
___ Advanced: Ex. 33–40, 42–48 even, 50–56, 57–72 multiples of 3, 75–81

Reteaching the Lesson
___ Extra Practice Copymasters page 15
___ Reteaching Copymasters page 15

Extending the Lesson
___ Applications Handbook pages 55–59

Notes

Lesson Plan 2.8
pages 107–113

2 days for Basic Course, 1 day for Full Course

Teacher's Name ______________________ *Class* __________ *Date* ________ *Room* ________

Goals 1. Use rates to relate quantities measured in different units.
2. Use ratios to relate quantities measured in the same units.

State/Local Objectives ____

NCTM Curriculum Standards: Problem Solving, Communication, Reasoning, Connections, Structure, Technology

✓ **Check items you wish to use for this lesson.**

Introducing the Lesson
___ Teaching Tools: Problem of the Day copymasters page 26, or Teacher's Edition page 107
___ Teaching Tools: Warm-Up Exercises copymasters page 56, or Teacher's Edition page 107

Teaching the Lesson using the following:
___ Color Transparencies page 12

Notes for substitute teacher ____

Closing the Lesson
___ Communicating about Algebra, Student's Edition page 109
___ Extend Communicating: Cooperative Learning, Teacher's Edition page 109
___ Guided Practice Exercises, Student's Edition page 110

Homework Assignment, pages 110–113
___ Basic/Average: Ex. 5–31 odd, 34, 37–44, 45–54 multiples of 3, 57
___ Above Average: Ex. 5–31 odd, 35–44, 49–59 odd
___ Advanced: Ex. 6–34 even, 35–44, 54–59

Reteaching the Lesson
___ Extra Practice Copymasters page 16
___ Reteaching Copymasters page 16

Extending the Lesson
___ Cultural Diversity Extensions page 2
___ Writing, Math Journal, Teacher's Edition page 109

Notes

Lesson Plan 3.1
pages 122–128

1 day for Basic Course, 1 day for Full Course

*Teacher's Name*__________ *Class*______ *Date*______ *Room*______

Goals
1. Solve equations systematically using addition, subtraction, and division.
2. Use reciprocals to solve equations.

State/Local Objectives __________

NCTM Curriculum Standards: Problem Solving, Communication, Connections, Reasoning, Geometry

✓ **Check items you wish to use for this lesson.**

Introducing the Lesson
___ Teaching Tools: Problem of the Day copymasters page 26, or Teacher's Edition page 122
___ Teaching Tools: Warm-Up Exercises copymasters page 56, or Teacher's Edition page 122

Notes for substitute teacher __________

Closing the Lesson
___ Communicating about Algebra, Student's Edition page 125
___ Guided Practice Exercises, Student's Edition page 126

Homework Assignment, pages 126–128
___ Basic/Average: Ex. 9–12, 13–51 odd, 55, 60–72 multiples of 3, 74, 75–79 odd
___ Above Average: Ex. 9–12, 14–24 even, 27–48 multiples of 3, 50–54 even, 56–57, 60–72 multiples of 3, 74, 75–80
___ Advanced: Ex. 9–12, 21–30, 43–48, 50–54, 56–58 even, 75–80

Reteaching the Lesson
___ Extra Practice Copymasters page 17
___ Reteaching Copymasters page 17
___ Alternative Assessment, Math Log pages 34–37

Extending the Lesson
___ Writing, Teacher's Edition page 128
___ Technology: Using Calculators and Computers page 20

Notes

Lesson Plan 3.2
pages 129–134

1 day for Basic Course, 1 day for Full Course

Teacher's Name ______________________ *Class* __________ *Date* __________ *Room* __________

Goal Use two or more transformations to solve an equation.

State/Local Objectives ____

NCTM Curriculum Standards: Problem Solving, Communication, Connections

✓ **Check items you wish to use for this lesson.**

Introducing the Lesson
___ Teaching Tools: Problem of the Day copymasters page 26, or Teacher's Edition page 129
___ Teaching Tools: Warm-Up Exercises copymasters page 57, or Teacher's Edition page 129

Teaching the Lesson using the following:
___ Alternative Approach, Teacher's Edition page 130
___ Alternative Approach, Teacher's Edition page 131

Notes for substitute teacher ____

Closing the Lesson
___ Communicating about Algebra, Student's Edition page 131
___ Guided Practice Exercises, Student's Edition page 132

Homework Assignment, pages 132–134
___ Basic/Average: Ex. 5–49 odd, 53–57 odd
___ Above Average: Ex. 12–33 multiples of 3, 36–40 even, 41–45, 51–57 odd
___ Advanced: Ex. 12–33 multiples of 3, 35, 39–42, 51–58

Reteaching the Lesson
___ Extra Practice Copymasters page 18
___ Reteaching Copymasters page 18

Extending the Lesson
___ Enrichment, Teacher's Edition page 134
___ Applications Handbook page 54
___ Cultural Diversity Extensions page 3
___ Technology: Using Calculators and Computers page 22

Notes

Lesson Plan 3.3
pages 135–140

1 day for Basic Course, 2 days for Full Course

Teacher's Name ____________ *Class* ______ *Date* ______ *Room* ______

Goals
1. Collect variables on one side of an equation.
2. Use algebraic models in real-life situations.

State/Local Objectives ________________________

NCTM Curriculum Standards: Problem Solving, Communication, Connections, Reasoning

✓ **Check items you wish to use for this lesson.**

Introducing the Lesson
___ Teaching Tools: Problem of the Day copymasters page 27, or Teacher's Edition page 135
___ Teaching Tools: Warm-Up Exercises copymasters page 57, or Teacher's Edition page 135

Teaching the Lesson using the following:
___ Common-Error Alert, Teacher's Edition page 136

Notes for substitute teacher ________________________

Closing the Lesson
___ Communicating about Algebra, Student's Edition page 137
___ Extend Communicating: Using Flowcharts, Teacher's Edition page 137
___ Guided Practice Exercises, Student's Edition page 138

Homework Assignment, pages 138–140
___ Basic/Average: Ex. 7–12, 15–36 multiples of 3, 37, 39–40, 45–59 odd
___ Above Average: Ex. 7–12, 21–35 odd, 37–38, 41–42, 45–54 multiples of 3, 55–63 odd
___ Advanced: Ex. 7–12, 31–40, 55–63

Reteaching the Lesson
___ Extra Practice Copymasters page 19
___ Reteaching Copymasters page 19

Extending the Lesson
___ Enrichment, Teacher's Edition page 140

Notes

Lesson Plan 3.4
pages 141–146

2 days for Basic Course, 2 days for Full Course

Teacher's Name ______________________ *Class* __________ *Date* __________ *Room* __________

Goal Use a problem-solving plan for problems that fit a linear model.

State/Local Objectives ____

NCTM Curriculum Standards: Problem Solving, Communication, Connections

✓ **Check items you wish to use for this lesson.**

Introducing the Lesson
___ Teaching Tools: Problem of the Day copymasters page 27, or Teacher's Edition page 141
___ Teaching Tools: Warm-Up Exercises copymasters page 57, or Teacher's Edition page 141

Teaching the Lesson using the following:
___ Teaching Tools: transparencies page 3

Notes for substitute teacher ____

Closing the Lesson
___ Communicating about Algebra, Student's Edition page 143
___ Extend Communicating: Writing, Teacher's Edition page 143
___ Guided Practice Exercises, Student's Edition page 144

Homework Assignment, pages 144–146
___ Basic/Average: Ex. 5–12, 18–25, 27–33 odd
___ Above Average: Ex. 5–8, 13–21, 26–32 even, 34–37
___ Advanced: Ex. 1–17, 22–25, 27–33 multiples of 3, 34–37

Reteaching the Lesson
___ Extra Practice Copymasters page 20
___ Reteaching Copymasters page 20

Notes

Lesson Plan 3.5
pages 148–153

1 day for Basic Course, 1 day for Full Course

Teacher's Name ______________________ *Class* __________ *Date* __________ *Room* ______

Goals 1. Find exact and approximate solutions of equations with decimals.
2. Solve problems that use decimal measurements.

State/Local Objectives ____

NCTM Curriculum Standards: Problem Solving, Communication, Connections

✓ **Check items you wish to use for this lesson.**

Introducing the Lesson
___ Teaching Tools: Problem of the Day copymasters page 27, or Teacher's Edition page 148
___ Teaching Tools: Warm-Up Exercises copymasters page 58, or Teacher's Edition page 148

Teaching the Lesson using the following:
___ Technology for Example 2, Teacher's Edition page 149
___ Color Transparencies page 15

Notes for substitute teacher ____

Closing the Lesson
___ Communicating about Algebra, Student's Edition page 150
___ Guided Practice Exercises, Student's Edition page 151

Homework Assignment, pages 151–153
___ Basic/Average: Ex. 5–15 odd, 18–39 multiples of 3, 41–42, 47–48, 49–57 multiples of 3, 58, 59
___ Above Average: Ex. 6–36 multiples of 3, 41–46, 57–59, 60–63
___ Advanced: Ex. 6–36 multiples of 3, 37–40, 43–48, 57–65

Reteaching the Lesson
___ Extra Practice Copymasters page 21
___ Reteaching Copymasters page 21

Extending the Lesson
___ Project, Teacher's Edition page 152
___ Enrichment, Teacher's Edition page 153

Notes

Lesson Plan 3.6
pages 154–159

2 days for Basic Course, 1 day for Full Course

Teacher's Name ______________________ *Class* ________ *Date* ________ *Room* ________

Goal Solve literal equations for a specified variable.

State/Local Objectives ______________________________

NCTM Curriculum Standards: Problem Solving, Communication, Connections, Geometry

✓ **Check items you wish to use for this lesson.**

Introducing the Lesson
___ Teaching Tools: Problem of the Day copymasters page 28, or Teacher's Edition page 154
___ Teaching Tools: Warm-Up Exercises copymasters page 58, or Teacher's Edition page 154

Teaching the Lesson using the following:
___ Alternative Approach for Examples 1–3, Teacher's Edition page 156
___ Color Transparencies page 16

Notes for substitute teacher ______________________________

Closing the Lesson
___ Communicating about Algebra, Student's Edition page 156
___ Extend Communicating: Cooperative Learning, Teacher's Edition page 157
___ Guided Practice Exercises, Student's Edition page 157

Homework Assignment, pages 157–158
___ Basic/Average: Ex. 5–33 odd, 35–38
___ Above Average: Ex. 5–33 odd, 35–38
___ Advanced: Ex. 6–24 even, 25–33 odd

Reteaching the Lesson
___ Extra Practice Copymasters page 22
___ Reteaching Copymasters page 22

Extending the Lesson
___ Writing, Teacher's Edition page 156
___ Enrichment, Teacher's Edition page 159
___ Technology: Using Calculators and Computers page 27

Notes

Lesson Plan 3.7
pages 160–165

1 day for Basic Course, 1 day for Full Course

Teacher's Name ______________________ *Class* __________ *Date* __________ *Room* __________

Goals
1. Use a coordinate plane to match points with ordered pairs of numbers.
2. Use a scatter plot.

State/Local Objectives ____

NCTM Curriculum Standards: Problem Solving, Communication, Connections, Statistics, Discrete Mathematics, Geometry

✓ **Check items you wish to use for this lesson.**

Introducing the Lesson
___ Teaching Tools: Problem of the Day copymasters page 28, or Teacher's Edition page 160
___ Teaching Tools: Warm-Up Exercises copymasters page 58, or Teacher's Edition page 160

Teaching the Lesson using the following:
___ Common-Error Alert, Teacher's Edition page 163
___ Teaching Tools: transparencies pages 1, 2, 4

Notes for substitute teacher ____

Closing the Lesson
___ Communicating about Algebra, Student's Edition page 162
___ Guided Practice Exercises, Student's Edition page 163

Homework Assignment, pages 163–165
___ Basic/Average: Ex. 7–11 odd, 13–18, 23–24, 27–35 odd, 36–37
___ Above Average: Ex. 7–11 odd, 19–22, 25–26, 31–34, 36–37
___ Advanced: Ex. 19–22, 23–26, 31–34, 36–37

Reteaching the Lesson
___ Extra Practice Copymasters page 23
___ Reteaching Copymasters page 23

Extending the Lesson
___ Enrichment, Teacher's Edition page 165
___ Writing, Teacher's Edition page 162
___ Applications Handbook, page 54
___ Technology: Using Calculators and Computers page 31

Notes

Lesson Plan 4.1
pages 176–182

1 day for Basic Course, 1 day for Full Course

Teacher's Name ______________________ *Class* __________ *Date* _________ *Room* ________

Goals

1. Graph horizontal and vertical lines.
2. Use equations of horizontal and vertical lines in real-life settings.

State/Local Objectives ____

NCTM Curriculum Standards: Problem Solving, Communication, Connections, Statistics, Geometry

✓ **Check items you wish to use for this lesson.**

Introducing the Lesson
___ Teaching Tools: Problem of the Day copymasters page 28, or Teacher's Edition page 176
___ Teaching Tools: Warm-Up Exercises copymasters page 59, or Teacher's Edition page 176

Teaching the Lesson using the following:
___ Common-Error Alert, Teacher's Edition page 177
___ Color Transparencies pages 18, 19
___ Teaching Tools: transparencies page 4

Notes for substitute teacher ____

Closing the Lesson
___ Communicating about Algebra, Student's Edition page 179
___ Extend Communicating: Step Functions, Teacher's Edition page 179
___ Guided Practice Exercises, Student's Edition page 180

Homework Assignment, pages 180–182
___ Basic/Average: Ex. 7–22, 23–29 odd, 31–33, 45–51 multiples of 3, 53–54
___ Above Average: Ex. 7–22, 34–42, 44–52 even, 53–54
___ Advanced: Ex. 7–22, 31–42, 47–51 odd, 53–54

Reteaching the Lesson
___ Extra Practice Copymasters page 24
___ Reteaching Copymasters page 24
___ Alternative Assessment: Math Log pages 38–40

Extending the Lesson
___ Historical Note, Teacher's Edition page 181
___ Connections, Teacher's Edition page 181
___ Enrichment, Teacher's Edition page 182
___ Technology: Using Calculators and Computers page 33

Notes

Lesson Plan 4.2
pages 183–188

1 day for Basic Course, 1 day for Full Course

Teacher's Name ______________________ *Class* __________ *Date* _________ *Room* ________

Goals

1. Graph a linear equation from a table of values.
2. Interpret graphs of linear equations.

State/Local Objectives ____

____

NCTM Curriculum Standards: Problem Solving, Communication, Connections, Statistics, Reasoning, Geometry

✓ **Check items you wish to use for this lesson.**

Introducing the Lesson
___ Teaching Tools: Problem of the Day copymasters page 29, or Teacher's Edition page 183
___ Teaching Tools: Warm-Up Exercises copymasters page 59, or Teacher's Edition page 183

Teaching the Lesson using the following:
___ Teaching Tools: transparencies pages 1, 2, 4

Notes for substitute teacher ____

____

____

Closing the Lesson
___ Communicating about Algebra, Student's Edition page 185
___ Extend Communicating: An Introduction to Slope, Teacher's Edition page 185
___ Guided Practice Exercises, Student's Edition page 186

Homework Assignment, pages 186–188
___ Basic/Average: Ex. 5–8, 9–19 odd, 21–26, 29–31, 33–43 odd, 45–47 odd
___ Above Average: Ex. 5–8, 19–25 odd, 27–30, 32–34, 45–47
___ Advanced: Ex. 5–8, 9–23 odd, 25–27, 29–31, 45–48

Reteaching the Lesson
___ Extra Practice Copymasters page 25
___ Reteaching Copymasters page 25

Extending the Lesson
___ Writing, Teacher's Edition page 186
___ Applications Handbook page 54

Notes

____

____

____

Lesson Plan 4.3
pages 190–195

1 day for Basic Course, 1 day for Full Course

Teacher's Name ______________________ *Class* __________ *Date* __________ *Room* __________

Goals
1. Find the intercepts of the graph of a linear equation.
2. Use intercepts to sketch a quick graph of a line.

State/Local Objectives ______________________________

NCTM Curriculum Standards: Problem Solving, Communication, Connections, Geometry

✓ **Check items you wish to use for this lesson.**

Introducing the Lesson
___ Teaching Tools: Problem of the Day copymasters page 29, or Teacher's Edition page 190
___ Teaching Tools: Warm-Up Exercises copymasters page 59, or Teacher's Edition page 190

Teaching the Lesson using the following:
___ Extension for Example 3, Teacher's Edition page 191
___ Teaching Tools: transparencies page 4

Notes for substitute teacher ______________________________

Closing the Lesson
___ Communicating about Algebra, Student's Edition page 193
___ Extend Communicating: Cooperative Learning, Teacher's Edition page 193
___ Guided Practice Exercises, Student's Edition page 193

Homework Assignment, pages 193–195
___ Basic/Average: Ex. 7–20, 21–47 odd, 49–50, 53–56, 59–60
___ Above Average: Ex. 7–20, 21–48 multiples of 3, 49–52, 54–58 even, 59–60, 63–64
___ Advanced: Ex. 7–14, 15–48 multiples of 3, 49–52, 59–64

Reteaching the Lesson
___ Extra Practice Copymasters page 26
___ Reteaching Copymasters page 26

Extending the Lesson
___ Writing, Teacher's Edition page 194
___ Writing, Teacher's Edition page 195

Notes

Lesson Plan 4.4
pages 197–203

1 day for Basic Course, 1 day for Full Course

Teacher's Name ______________________ *Class* __________ *Date* __________ *Room* __________

Goals

1. Find the slope of a line using two of its points.
2. Interpret slope as constant or average rate of change.

State/Local Objectives ____

____

NCTM Curriculum Standards: Problem Solving, Communication, Connections, Statistics, Geometry, Reasoning, Calculus

✓ **Check items you wish to use for this lesson.**

Introducing the Lesson
___ Teaching Tools: Problem of the Day copymasters page 29, or Teacher's Edition page 197
___ Teaching Tools: Warm-Up Exercises copymasters page 60, or Teacher's Edition page 197

Teaching the Lesson using the following:
___ Common-Error Alert, Teacher's Edition page 198
___ Common-Error Alert, Teacher's Edition page 199
___ Alternative Approach: Using Graphing Calculators, Teacher's Edition page 199
___ Cooperative Learning: Example 5, Teacher's Edition page 200
___ Color Transparencies pages 19, 20

Notes for substitute teacher ____

____

____

Closing the Lesson
___ Communicating about Algebra, Student's Edition page 200
___ Guided Practice Exercises, Student's Edition page 201

Homework Assignment, pages 201–203
___ Basic/Average: Ex. 9–16, 17–37 odd, 43, 45, 47–50, 51–59 odd
___ Above Average: Ex. 9–16, 18–34 even, 35–42, 44, 46, 51–57 multiples of 3, 59–62
___ Advanced: Ex. 9–36 multiples of 3, 39–42, 45, 46–50, 59–63

Reteaching the Lesson
___ Extra Practice Copymasters page 27
___ Reteaching Copymasters page 27

Extending the Lesson
___ Enrichment, Teacher's Edition page 202
___ Writing, Teacher's Edition page 202
___ Cultural Diversity Extensions page 4
___ Technology: Using Calculators and Computers page 35

Notes

____

____

Lesson Plan 4.5
pages 205–210

1 day for Basic Course, 1 day for Full Course

Teacher's Name ______________________ *Class* __________ *Date* __________ *Room* __________

Goals
1. Find the slope and y-intercept from an equation.
2. Use the slope-intercept form to sketch a line and solve problems.

State/Local Objectives ______________________________

NCTM Curriculum Standards: Problem Solving, Communication, Connections, Geometry, Reasoning

✓ **Check items you wish to use for this lesson.**

Introducing the Lesson
___ Teaching Tools: Problem of the Day copymasters page 30, or Teacher's Edition page 205
___ Teaching Tools: Warm-Up Exercises copymasters page 60, or Teacher's Edition page 205

Teaching the Lesson using the following:
___ Common-Error Alert, Teacher's Edition page 207
___ Common-Error Alert, Teacher's Edition page 208

Notes for substitute teacher ______________________________

Closing the Lesson
___ Communicating about Algebra, Student's Edition page 207
___ Guided Practice Exercises, Student's Edition page 208

Homework Assignment, pages 208–210
___ Basic/Average: Ex. 5–27 odd, 29–32, 33–43 odd, 45–54 multiples of 3, 58, 59
___ Above Average: Ex. 6–27 multiples of 3, 29–32, 36, 39–40, 42–43, 45–54 multiples of 3, 57–62
___ Advanced: Ex. 6–27 multiples of 3, 29–32, 37–40, 42–43, 57–62

Reteaching the Lesson
___ Extra Practice Copymasters page 28
___ Reteaching Copymasters page 28

Extending the Lesson
___ Writing, Teacher's Edition page 209
___ Technology: Using Calculators and Computers page 37

Notes

Lesson Plan 4.6
pages 211–216

1 day for Basic Course, 1 day for Full Course

Teacher's Name ______________________ *Class* __________ *Date* __________ *Room* __________

Goals

1. Use a graph as a quick check of a solution found algebraically.
2. Approximate solutions of real-life problems by using a graph.

State/Local Objectives ____

NCTM Curriculum Standards: Problem Solving, Communication, Connections, Reasoning, Geometry, Technology

✓ **Check items you wish to use for this lesson.**

Introducing the Lesson
___ Teaching Tools: Problem of the Day copymasters page 30, or Teacher's Edition page 211
___ Teaching Tools: Warm-Up Exercises copymasters page 60, or Teacher's Edition page 211

Teaching the Lesson using the following:
___ Technology for Example 2, Teacher's Edition page 212

Notes for substitute teacher ____

Closing the Lesson
___ Communicating about Algebra, Student's Edition page 213
___ Guided Practice Exercises, Student's Edition page 214

Homework Assignment, pages 214–215
___ Basic/Average: Ex. 5–8, 9, 11, 15–24 multiples of 3, 25–31 odd, 33–42 multiples of 3, 43–46
___ Above Average: Ex. 5–8, 10, 12, 13–23 odd, 26–32 even, 42–46
___ Advanced: Ex. 5–8, 11–12, 19–24, 25–29 odd, 31–32, 43–46

Reteaching the Lesson
___ Extra Practice Copymasters page 29
___ Reteaching Copymasters page 29

Extending the Lesson
___ Enrichment, Teacher's Edition page 216
___ Technology: Using Calculators and Computers page 40

Notes

Lesson Plan 4.7
pages 217–222

3 days for Basic Course, 3 days for Full Course

Teacher's Name ______________________ *Class* __________ *Date* __________ *Room* __________

Goals
1. Graph an absolute value equation.
2. Model a real-life situation using graphs of absolute value equations.

State/Local Objectives ____

NCTM Curriculum Standards: Problem Solving, Communication, Reasoning, Connections, Technology, Geometry

✓ **Check items you wish to use for this lesson.**

Introducing the Lesson
___ Teaching Tools: Problem of the Day copymasters page 30, or Teacher's Edition page 217
___ Teaching Tools: Warm-Up Exercises copymasters page 61, or Teacher's Edition page 217

Teaching the Lesson using the following:
___ Connections for Example 3, Teacher's Edition page 219
___ Color Transparencies pages 20, 21
___ Teaching Tools: transparencies page 10

Notes for substitute teacher ____

Closing the Lesson
___ Communicating about Algebra, Student's Edition page 219
___ Guided Practice Exercises, Student's Edition page 220

Homework Assignment, pages 220–222
___ Basic/Average: Ex. 5–31 odd, 33–51 multiples of 3, 55–61 odd, 62
___ Above Average: Ex. 5–31 odd, 33–51 multiples of 3, 52, 54, 55–63 odd
___ Advanced: Ex. 5–31 odd, 33–51 multiples of 3, 52–54, 56–60 even, 62–63

Reteaching the Lesson
___ Extra Practice Copymasters page 30
___ Reteaching Copymasters page 30

Extending the Lesson
___ Technology: Using Calculators and Computers page 42

Notes

Lesson Plan 4.8
pages 223–228

3 days for Basic Course, 3 days for Full Course

Teacher's Name ______________________ *Class* __________ *Date* ________ *Room* ________

Goals
1. Solve and check absolute value equations algebraically.
2. Use a graph to check solutions of absolute value equations.

State/Local Objectives ____

NCTM Curriculum Standards: Problem Solving, Communication, Reasoning, Connections, Geometry

✓ **Check items you wish to use for this lesson.**

Introducing the Lesson
___ Teaching Tools: Problem of the Day copymasters page 31, or Teacher's Edition page 223
___ Teaching Tools: Warm-Up Exercises copymasters page 61, or Teacher's Edition page 223

Teaching the Lesson using the following:
___ Color Transparencies page 21
___ Teaching Tools: transparencies pages 4, 10

Notes for substitute teacher ____

Closing the Lesson
___ Communicating about Algebra, Student's Edition page 225
___ Extend Communicating: Writing, Teacher's Edition page 225
___ Guided Practice Exercises, Student's Edition page 226

Homework Assignment, pages 226–227
___ Basic/Average: Ex. 1–33 odd, 38–39, 40–50 even, 52–53
___ Above Average: Ex. 1–4, 6–33 multiples of 3, 34–38, 41–51 odd, 52–53, 54, 56
___ Advanced: Ex. 1–4, 6–33 multiples of 3, 34–39, 41–51 odd, 52–57

Reteaching the Lesson
___ Extra Practice Copymasters page 31
___ Reteaching Copymasters page 31

Extending the Lesson
___ Technology: Teacher's Edition page 227
___ Writing, Teacher's Edition page 227

Notes

Lesson Plan 5.1
pages 236–241

1 day for Basic Course, 1 day for Full Course

*Teacher's Name*__________ *Class*______ *Date*______ *Room*______

Goals
1. Use the slope-intercept form to write an equation of a line.
2. Model a real-life situation with a linear equation.

State/Local Objectives ____________________

NCTM Curriculum Standards: Problem Solving, Communication, Statistics, Connections

✓ **Check items you wish to use for this lesson.**

Introducing the Lesson
___ Teaching Tools: Problem of the Day copymasters page 31, or Teacher's Edition page 236
___ Teaching Tools: Warm-Up Exercises copymasters page 61, or Teacher's Edition page 236

Notes for substitute teacher ____________________

Closing the Lesson
___ Communicating about Algebra, Student's Edition page 238
___ Guided Practice Exercises, Student's Edition page 239

Homework Assignment, pages 239–240
___ Basic/Average: Ex. 9–19 odd, 25–26, 31–41 odd, 47–48
___ Above Average: Ex. 10–22 even, 24, 27–28, 31–45 odd, 47–48 odd, 49–52
___ Advanced: Ex. 10–20 even, 25–28, 32–46 even, 47–52

Reteaching the Lesson
___ Extra Practice Copymasters page 32
___ Reteaching Copymasters page 32
___ Alternative Assessment: Math Log pages 41–43

Extending the Lesson
___ Research Project, Teacher's Edition page 240
___ Enrichment, Teacher's Edition page 241

Notes

Lesson Plan 5.2
pages 242–247

2 days for Basic Course, 1 day for Full Course

Teacher's Name ______________________ *Class* __________ *Date* __________ *Room* __________

Goals
1. Use the slope and any point on the line to write an equation of the line.
2. Model a real-life situation with a linear equation.

State/Local Objectives ______________________

NCTM Curriculum Standards: Problem Solving, Communication, Connections, Reasoning

✓ **Check items you wish to use for this lesson.**

Introducing the Lesson
___ Teaching Tools: Problem of the Day copymasters page 31, or Teacher's Edition page 242
___ Teaching Tools: Warm-Up Exercises copymasters page 62, or Teacher's Edition page 242

Notes for substitute teacher ______________________

Closing the Lesson
___ Communicating about Algebra, Student's Edition page 244
___ Extend Communicating: Cooperative Learning, Teacher's Edition page 244
___ Guided Practice Exercises, Student's Edition page 245

Homework Assignment, pages 245–247
___ Basic/Average: Ex. 9–27 odd, 29, 31, 35, 37–45 odd, 47–49
___ Above Average: Ex. 9–27 odd, 31–35, 39–45 multiples of 3, 47–51
___ Advanced: Ex. 9–19 odd, 20–32 even, 33–36, 39–45 odd, 47–51

Reteaching the Lesson
___ Extra Practice Copymasters page 33
___ Reteaching Copymasters page 33

Extending the Lesson
___ Enrichment, Teacher's Edition page 246

Notes

Lesson Plan 5.3
pages 249–254

1 day for Basic Course, 1 day for Full Course

Teacher's Name ________ *Class* ________ *Date* ________ *Room* ________

Goals
1. Write an equation of a line given two points on the line.
2. Model a real-life situation with a linear equation.

State/Local Objectives ________

NCTM Curriculum Standards: Problem Solving, Communication, Connections

✓ **Check items you wish to use for this lesson.**

Introducing the Lesson
___ Teaching Tools: Problem of the Day copymasters page 32, or Teacher's Edition page 249
___ Teaching Tools: Warm-Up Exercises copymasters page 62, or Teacher's Edition page 249

Teaching the Lesson using the following:
___ Common-Error Alert, Teacher's Edition page 249
___ Color Transparencies page 24

Notes for substitute teacher ________

Closing the Lesson
___ Communicating about Algebra, Student's Edition page 251
___ Guided Practice Exercises, Student's Edition page 252

Homework Assignment, pages 252–254
___ Basic/Average: Ex. 7–10, 11–25 odd, 27–28, 31–41 odd, 45–46
___ Above Average: Ex. 8–26 even, 29–30, 32–46 even
___ Advanced: Ex. 8–26 even, 27–30, 33–34, 36–45 multiples of 3, 46–47

Reteaching the Lesson
___ Extra Practice Copymasters page 34
___ Reteaching Copymasters page 34

Extending the Lesson
___ Enrichment, Teacher's Edition page 253
___ Cultural Diversity Extensions page 5

Notes

Lesson Plan 5.4
pages 255–261

1 day for Basic Course, 1 day for Full Course

Teacher's Name ______________________ *Class* __________ *Date* __________ *Room* __________

Goals
1. Transform a linear equation into standard form.
2. Model a real-life situation using the standard form of a linear equation.

State/Local Objectives ______________________________

____

NCTM Curriculum Standards: Problem Solving, Communication, Statistics, Reasoning, Discrete Mathematics

✓ **Check items you wish to use for this lesson.**

Introducing the Lesson
___ Teaching Tools: Problem of the Day copymasters page 32, or Teacher's Edition page 255
___ Teaching Tools: Warm-Up Exercises copymasters page 62, or Teacher's Edition page 255

Teaching the Lesson using the following:
___ Common-Error Alert, Teacher's Edition page 256
___ Color Transparencies pages 24, 25
___ Teaching Tools: transparencies pages 1, 2, 4

Notes for substitute teacher ______________________________

____

____

Closing the Lesson
___ Communicating about Algebra, Student's Edition page 258
___ Extend Communicating: Research, Teacher's Edition page 258
___ Guided Practice Exercises, Student's Edition page 259

Homework Assignment, pages 259–261
___ Basic/Average: Ex. 7–10, 11–27 odd, 28–29
___ Above Average: Ex. 7–10, 12–26 even, 28–31
___ Advanced: Ex. 7–10, 12–16 even, 17–19, 21–27 odd, 28–31

Reteaching the Lesson
___ Extra Practice Copymasters page 35
___ Reteaching Copymasters page 35

Extending the Lesson
___ Enrichment, Teacher's Edition page 260
___ Technology: Using Calculators and Computers pages 45, 48, 51

Notes

____

____

____

Lesson Plan 5.5
pages 264–269

1 day for Basic Course, 1 day for Full Course

Teacher's Name ____________ *Class* ______ *Date* ______ *Room* ______

Goals
1. Transform a linear equation into standard form.
2. Model a real-life situation using the standard form of an equation.

State/Local Objectives ____________

NCTM Curriculum Standards: Problem Solving, Communications, Connections, Statistics

✓ **Check items you wish to use for this lesson.**

Introducing the Lesson
___ Teaching Tools: Problem of the Day copymasters page 32, or Teacher's Edition page 264
___ Teaching Tools: Warm-Up Exercises copymasters page 63, or Teacher's Edition page 264

Teaching the Lesson using the following:
___ Color Transparencies pages 26
___ Teaching Tools: transparencies page 4

Notes for substitute teacher ____________

Closing the Lesson
___ Communicating about Algebra, Student's Edition page 266
___ Extend Communicating, Teacher's Edition page 266
___ Guided Practice Exercises, Student's Edition page 267

Homework Assignment, pages 267–268
___ Basic/Average: Ex. 7–17 odd, 19–22, 25, 27, 29–32, 35–43 odd, 45–47
___ Above Average: Ex. 10–18 even, 19–20, 23–24, 27–32, 36–44 even, 45–49
___ Advanced: Ex. 9–18 multiples of 3, 19–27 odd, 31–34, 36–44 even, 45–49

Reteaching the Lesson
___ Extra Practice Copymasters page 36
___ Reteaching Copymasters page 36

Extending the Lesson
___ Writing, Teacher's Edition page 266
___ Enrichment, Teacher's Edition page 269

Notes

Lesson Plan 5.6
pages 270–276

1 day for Basic Course, 1 day for Full Course

Teacher's Name ______________________ *Class* __________ *Date* __________ *Room* __________

Goals

1. Use the point-slope form to write a linear equation.
2. Model a real-life situation using the point-slope form.

State/Local Objectives ______________________

NCTM Curriculum Standards: Problem Solving, Communication, Connections, Reasoning, Geometry

✓ **Check items you wish to use for this lesson.**

Introducing the Lesson
___ Teaching Tools: Problem of the Day copymasters page 33, or Teacher's Edition page 270
___ Teaching Tools: Warm-Up Exercises copymasters page 63, or Teacher's Edition page 270

Teaching the Lesson using the following:
___ Common-Error Alert, Teacher's Edition page 270
___ Common-Error Alert, Teacher's Edition page 271
___ Color Transparencies page 26

Notes for substitute teacher ______________________

Closing the Lesson
___ Communicating about Algebra, Student's Edition page 273
___ Extend Communicating: An Experiment, Teacher's Edition page 273
___ Guided Practice Exercises, Student's Edition page 274

Homework Assignment, pages 274–275
___ Basic/Average: Ex. 9–17 odd, 20, 21–25 odd, 27–32
___ Above Average: Ex. 10–26 even, 27–32
___ Advanced: Ex. 10–26 even, 27–32

Reteaching the Lesson
___ Extra Practice Copymasters page 37
___ Reteaching Copymasters page 37

Extending the Lesson
___ Enrichment, Teacher's Edition page 275

Notes

Lesson Plan 5.7
pages 277–283

2 days for Basic Course, 1 day for Full Course

Teacher's Name ______________________ *Class* __________ *Date* __________ *Room* __________

Goals
1. Create and use linear models to solve problems.
2. Make accurate and easy-to-use linear models.

State/Local Objectives ______________________________

____

NCTM Curriculum Standards: Problem Solving, Communication, Connections, Statistics, Discrete Mathematics, Geometry

✓ **Check items you wish to use for this lesson.**

Introducing the Lesson
___ Teaching Tools: Problem of the Day copymasters page 33, or Teacher's Edition page 277
___ Teaching Tools: Warm-Up Exercises copymasters page 63, or Teacher's Edition page 277

Teaching the Lesson using the following:
___ Extension for Example 4, Teacher's Edition page 278
___ Color Transparencies page 27

Notes for substitute teacher ______________________________

____

____

Closing the Lesson
___ Communicating about Algebra, Student's Edition page 280
___ Extend Communicating, Teacher's Edition page 280
___ Guided Practice Exercises, Student's Edition page 281

Homework Assignment, pages 281–282
___ Basic/Average: Ex. 7–12, 19–22, 23–26, 29–31
___ Above Average: Ex. 7–16, 27–31, 32–35
___ Advanced: Ex. 17–22, 27–35

Reteaching the Lesson
___ Extra Practice Copymasters page 38
___ Reteaching Copymasters page 38

Extending the Lesson
___ Enrichment, Teacher's Edition page 283

Notes

____

____

____

Lesson Plan 6.1
pages 292–297

1 day for Basic Course, 1 day for Full Course

Teacher's Name ______________ *Class* ______ *Date* ______ *Room* ______

Goals
1. Graph linear inequalities in one variable.
2. Solve linear inequalities in one variable.

State/Local Objectives ______________

NCTM Curriculum Standards: Problem Solving, Communication, Connections, Structure, Reasoning, Geometry

✓ **Check items you wish to use for this lesson.**

Introducing the Lesson
___ Teaching Tools: Problem of the Day copymasters page 33, or Teacher's Edition page 292
___ Teaching Tools: Warm-Up Exercises copymasters page 64, or Teacher's Edition page 292

Teaching the Lesson using the following:
___ Common-Error Alert, Teacher's Edition page 293
___ Teaching Tools: transparencies page 6

Notes for substitute teacher ______________

Closing the Lesson
___ Communicating about Algebra, Student's Edition page 295
___ Guided Practice Exercises, Student's Edition page 296

Homework Assignment, pages 296–297
___ Basic/Average: Ex. 9–31 odd, 33–41, 45–46, 47–53 odd, 55
___ Above Average: Ex. 10–32 even, 33–41, 42–46 even, 47–53 odd, 55–56
___ Advanced: Ex. 9–30 multiples of 3, 33–41, 43–46, 49–53 odd, 55–56

Reteaching the Lesson
___ Extra Practice Copymasters page 39
___ Reteaching Copymasters page 39
___ Alternative Assessment: Math Log pages 44–46

Extending the Lesson
___ Writing, Teacher's Edition page 295
___ Enrichment, Teacher's Edition page 297

Notes

Lesson Plan 6.2
pages 298–303

3 days for Basic Course, 3 days for Full Course

Teacher's Name ______________ *Class* ________ *Date* ________ *Room* ________

Goal Write and use a linear inequality as a model for a real-life situation.

State/Local Objectives ______________

NCTM Curriculum Standards: Problem Solving, Communication, Connections, Reasoning, Geometry

✓ **Check items you wish to use for this lesson.**

Introducing the Lesson
___ Teaching Tools: Problem of the Day copymasters page 34, or Teacher's Edition page 298
___ Teaching Tools: Warm-Up Exercises copymasters page 64, or Teacher's Edition page 298

Teaching the Lesson using the following:
___ Teaching Tools: transparencies pages 3, 4

Notes for substitute teacher ______________

Closing the Lesson
___ Communicating about Algebra, Student's Edition page 300
___ Extend Communicating: Writing Problems, Cooperative Learning, Teacher's Edition page 300
___ Guided Practice Exercises, Student's Edition page 301

Homework Assignment, pages 301–303
___ Basic/Average: Ex. 5–11 odd, 15–27 odd, 29–32
___ Above Average: Ex. 6–16 even, 23–27 odd, 29–32
___ Advanced: Ex. 6–12 multiples of 3, 14–17, 20–26 even, 27–28, 29–32

Reteaching the Lesson
___ Extra Practice Copymasters page 40
___ Reteaching Copymasters page 40

Extending the Lesson
___ Enrichment, Teacher's Edition page 303

Notes

Lesson Plan 6.3
pages 305–310

3 days for Basic Course, 3 days for Full Course

Teacher's Name ______________________ *Class* __________ *Date* __________ *Room* __________

Goals
1. Solve and graph compound inequalities.
2. Model a real-life situation with a compound inequality.

State/Local Objectives ____

NCTM Curriculum Standards: Problem Solving, Communication, Connections, Reasoning, Geometry

✓ **Check items you wish to use for this lesson.**

Introducing the Lesson
___ Teaching Tools: Problem of the Day copymasters page 34, or Teacher's Edition page 305
___ Teaching Tools: Warm-Up Exercises copymasters page 64, or Teacher's Edition page 305

Teaching the Lesson using the following:
___ Alternative Approach for Example 2, Teacher's Edition page 306
___ Color Transparencies page 28
___ Teaching Tools: transparencies pages 1, 2, 4, 6

Notes for substitute teacher ____

Closing the Lesson
___ Communicating about Algebra, Student's Edition page 307
___ Guided Practice Exercises, Student's Edition page 308

Homework Assignment, pages 308–310
___ Basic/Average: Ex. 1–6, 9–27 multiples of 3, 29–30, 33, 35, 37–43 odd, 45
___ Above Average: Ex. 1–6, 11–14, 17–20, 21–24, 25–31 odd, 34–35, 36–42 multiples of 3, 45–46
___ Advanced: Ex. 1–6, 11–13, 17–20, 23–24, 26–35, 42–48

Reteaching the Lesson
___ Extra Practice Copymasters page 41
___ Reteaching Copymasters page 41

Extending the Lesson
___ History, Teacher's Edition page 309
___ Project, Teacher's Edition page 309
___ Enrichment, Teacher's Edition page 310
___ Technology page 310

Notes

Lesson Plan 6.4
pages 313–319

2 days for Basic Course, 2 days for Full Course

Teacher's Name ______________________ *Class* __________ *Date* __________ *Room* __________

Goals
1. Solve absolute value inequalities.
2. Model a real-life situation with an absolute value inequality.

State/Local Objectives ____

NCTM Curriculum Standards: Problem Solving, Communication, Connections, Reasoning, Geometry

✓ **Check items you wish to use for this lesson.**

Introducing the Lesson
___ Teaching Tools: Problem of the Day copymasters page 34, or Teacher's Edition page 313
___ Teaching Tools: Warm-Up Exercises copymasters page 65, or Teacher's Edition page 313

Teaching the Lesson using the following:
___ Color Transparencies page 29
___ Teaching Tools: transparencies pages 4, 6, 10

Notes for substitute teacher ____

Closing the Lesson
___ Communicating about Algebra, Student's Edition page 315
___ Guided Practice Exercises, Student's Edition page 316

Homework Assignment, pages 316–318
___ Basic/Average: Ex. 7–12, 19–37 odd, 41–45 odd, 48–57 multiples of 3, 59–65 odd
___ Above Average: Ex. 7–12, 20–38 even, 42–46 even, 51–54, 59–66
___ Advanced: Ex. 7–12, 15–36 multiples of 3, 39–42, 45–57 multiples of 3, 59–66

Reteaching the Lesson
___ Extra Practice Copymasters page 42
___ Reteaching Copymasters page 42

Extending the Lesson
___ Enrichment, Teacher's Edition page 318
___ Applications Handbook page 36
___ Technology: Using Calculators and Computers page 53

Notes

Lesson Plan 6.5
pages 320–325

2 days for Basic Course, 1 day for Full Course

Teacher's Name ______________________ *Class* __________ *Date* __________ *Room* __________

Goals
1. Graph a linear inequality in two variables.
2. Model a real-life situation using a linear inequality in two variables.

State/Local Objectives ____

____

NCTM Curriculum Standards: Problem Solving, Communication, Connections, Discrete Mathematics, Statistics, Reasoning

✓ **Check items you wish to use for this lesson.**

Introducing the Lesson
___ Teaching Tools: Problem of the Day copymasters page 35, or Teacher's Edition page 320
___ Teaching Tools: Warm-Up Exercises copymasters page 65, or Teacher's Edition page 320

Teaching the Lesson using the following:
___ Common-Error Alert, Teacher's Edition page 320
___ Color Transparencies page 30
___ Teaching Tools: transparencies pages 4, 11, 12

Notes for substitute teacher ____

____

____

Closing the Lesson
___ Communicating about Algebra, Student's Edition page 322
___ Guided Practice Exercises, Student's Edition page 323

Homework Assignment, pages 323–325
___ Basic/Average: Ex. 7–10, 11–14, 15–21 odd, 23–25, 27–45 multiples of 3, 47
___ Above Average: Ex. 8, 10, 11–14, 16–22 even, 24–25, 27–45 multiples of 3, 47–48
___ Advanced: Ex. 7, 9, 11–14, 16–22 even, 24–26, 31–34, 47–48

Reteaching the Lesson
___ Extra Practice Copymasters page 43
___ Reteaching Copymasters page 43

Extending the Lesson
___ Enrichment, Teacher's Edition page 325
___ Writing, Teacher's Edition page 325

Notes

____

____

____

Lesson Plan 6.6
pages 327–333

1 day for Basic Course, 1 day for Full Course

Teacher's Name ______________________ *Class* __________ *Date* __________ *Room* __________

Goal Draw and interpret visual models such as time lines, picture graphs, and circle graphs.

State/Local Objectives ____

____

NCTM Curriculum Standards: Problem Solving, Communication, Connections, Discrete Mathematics, Statistics, Reasoning

✓ **Check items you wish to use for this lesson.**

Introducing the Lesson
___ Teaching Tools: Problem of the Day copymasters page 35, or Teacher's Edition page 327
___ Teaching Tools: Warm-Up Exercises copymasters page 65, or Teacher's Edition page 327

Teaching the Lesson using the following:
___ Extension for Example 3, Teacher's Edition page 328
___ Color Transparencies pages 30–34
___ Teaching Tools: transparencies page 5

Notes for substitute teacher ____

____

____

Closing the Lesson
___ Communicating about Algebra, Student's Edition page 329
___ Guided Practice Exercises, Student's Edition page 330

Homework Assignment, pages 330–332
___ Basic/Average: Ex. 5–8, 11–15, 20–22, 25–39 odd, 40–41
___ Above Average: Ex. 5–10, 11, 16–19, 23–25, 26–39 multiples of 3, 40–42
___ Advanced: Ex. 11–19, 22–23, 27–36 multiples of 3, 40–42

Reteaching the Lesson
___ Extra Practice Copymasters page 44
___ Reteaching Copymasters page 44

Extending the Lesson
___ Writing, Teacher's Edition page 330
___ Enrichment, Teacher's Edition page 332
___ Cultural Diversity Extensions page 6

Notes

____

____

____

Lesson Plan 7.1
pages 346–351

1 day for Basic Course, 1 day for Full Course

Teacher's Name ______________________ *Class* __________ *Date* ________ *Room* ________

Goals
1. Solve a system of linear equations by graphing.
2. Model a real-life situation using a system of linear equations.

State/Local Objectives ____

NCTM Curriculum Standards: Problem Solving, Communication, Connections, Statistics, Discrete Mathematics, Geometry

✓ **Check items you wish to use for this lesson.**

Introducing the Lesson
___ Teaching Tools: Problem of the Day copymasters page 35, or Teacher's Edition page 346
___ Teaching Tools: Warm-Up Exercises copymasters page 66, or Teacher's Edition page 346

Teaching the Lesson using the following:
___ Color Transparencies page 36
___ Teaching Tools: transparencies pages 1, 2, 4

Notes for substitute teacher ____

Closing the Lesson
___ Communicating about Algebra, Student's Edition page 348
___ Extend Communicating: Writing, Teacher's Edition page 349
___ Guided Practice Exercises, Student's Edition page 349

Homework Assignment, pages 349–351
___ Basic/Average: Ex. 7–29 odd, 41–47 odd, 53–57 odd, 61–62
___ Above Average: Ex. 7–35 odd, 39–40, 41–59 odd, 61–62
___ Advanced: Ex. 13–35 odd, 37–39, 49–59 odd, 61–62

Reteaching the Lesson
___ Extra Practice Copymasters page 45
___ Reteaching Copymasters page 45
___ Alternative Assessment: Math Log pages 47–50

Extending the Lesson
___ Enrichment, Teacher's Edition page 351
___ Technology: Using Calculators and Computers page 55

Notes

Lesson Plan 7.2
pages 353–358

1 day for Basic Course, 1 day for Full Course

Teacher's Name ____________ *Class* ________ *Date* ________ *Room* ________

Goals

1. Use substitution to solve a linear system.
2. Model a real-life situation using a system of linear equations.

State/Local Objectives ____________

NCTM Curriculum Standards: Problem Solving, Communication, Connections, Reasoning, Discrete Mathematics, Geometry

✓ **Check items you wish to use for this lesson.**

Introducing the Lesson
- ___ Teaching Tools: Problem of the Day copymasters page 36, or Teacher's Edition page 353
- ___ Teaching Tools: Warm-Up Exercises copymasters page 66, or Teacher's Edition page 353

Teaching the Lesson using the following:
- ___ Color Transparencies page 37
- ___ Teaching Tools: transparencies page 4

Notes for substitute teacher ____________

Closing the Lesson
- ___ Communicating about Algebra, Student's Edition page 355
- ___ Extend Communicating, Teacher's Edition page 355
- ___ Guided Practice Exercises, Student's Edition page 356

Homework Assignment, pages 356–358
- ___ Basic/Average: Ex. 5–27 odd, 29, 35, 37–41 odd, 49, 53
- ___ Above Average: Ex. 5–27 odd, 29–37 odd, 45–53 odd, 54
- ___ Advanced: Ex. 9–27 odd, 30–36, 49, 51, 53, 54

Reteaching the Lesson
- ___ Extra Practice Copymasters page 46
- ___ Reteaching Copymasters page 46

Extending the Lesson
- ___ Writing, Teacher's Edition page 357
- ___ Technology, Teacher's Edition page 357
- ___ Enrichment, Teacher's Edition page 358

Notes

Lesson Plan 7.3

pages 360–366

1 day for Basic Course, 1 day for Full Course

Teacher's Name ______________________ *Class* __________ *Date* ________ *Room* ________

Goals

1. Use linear combinations to solve a linear system.
2. Model a real-life situation using a system of linear equations.

State/Local Objectives ____

NCTM Curriculum Standards: Problem Solving, Communication, Connections, Discrete Mathematics

✓ **Check items you wish to use for this lesson.**

Introducing the Lesson

___ Teaching Tools: Problem of the Day copymasters page 36, or Teacher's Edition page 360
___ Teaching Tools: Warm-Up Exercises copymasters page 66, or Teacher's Edition page 360

Teaching the Lesson using the following:

___ Color Transparencies page 37

Notes for substitute teacher ____

Closing the Lesson

___ Communicating about Algebra, Student's Edition page 362
___ Guided Practice Exercises, Student's Edition page 363

Homework Assignment, pages 363–365

___ Basic/Average: Ex. 7–23 odd, 33, 39–47 odd, 51, 53
___ Above Average: Ex. 7–27 odd, 29, 31, 33, 35–47 odd, 51–53
___ Advanced: Ex. 17–29 odd, 30, 32, 33–45 odd, 49–54

Reteaching the Lesson

___ Extra Practice Copymasters page 47
___ Reteaching Copymasters page 47

Extending the Lesson

___ Enrichment, Teacher's Edition page 366
___ Applications Handbook page 3
___ Technology: Using Calculators and Computers page 57

Notes

Lesson Plan 7.4
pages 367–372

2 days for Basic Course, 2 days for Full Course

Teacher's Name ____________ *Class* ______ *Date* ______ *Room* ______

Goal Write and use a linear system as a real-life model.

State/Local Objectives ____________

NCTM Curriculum Standards: Problem Solving, Communication, Connections, Discrete Mathematics, Geometry

✓ **Check items you wish to use for this lesson.**

Introducing the Lesson
___ Teaching Tools: Problem of the Day copymasters page 36, or Teacher's Edition page 367
___ Teaching Tools: Warm-Up Exercises copymasters page 67, or Teacher's Edition page 367

Teaching the Lesson using the following:
___ Teaching Tools: transparencies page 3

Notes for substitute teacher ____________

Closing the Lesson
___ Communicating about Algebra, Student's Edition page 369
___ Guided Practice Exercises, Student's Edition page 370

Homework Assignment, pages 370–371
___ Basic/Average: Ex. 5–11 odd, 15–27 odd, 31, 32
___ Above Average: Ex. 5–27 odd, 31–32
___ Advanced: Ex. 7–14, 28–33

Reteaching the Lesson
___ Extra Practice Copymasters page 48
___ Reteaching Copymasters page 48

Extending the Lesson
___ Enrichment, Teacher's Edition page 372
___ Applications Handbook page 54

Notes

Lesson Plan 7.5
pages 374–380

1 day for Basic Course, 1 day for Full Course

Teacher's Name ______________________ *Class* ________ *Date* ________ *Room* ________

Goals
1. Visualize the solution possibilities for linear systems.
2. Identify a linear system that has many solutions.

State/Local Objectives ______________________

NCTM Curriculum Standards: Problem Solving, Communication, Connections, Reasoning, Discrete Mathematics, Geometry

✓ **Check items you wish to use for this lesson.**

Introducing the Lesson
___ Teaching Tools: Problem of the Day copymasters page 37, or Teacher's Edition page 374
___ Teaching Tools: Warm-Up Exercises copymasters page 67, or Teacher's Edition page 374

Teaching the Lesson using the following:
___ Alternative Approach: Technology, Teacher's Edition page 375
___ Color Transparencies page 38
___ Teaching Tools: transparencies page 4

Notes for substitute teacher ______________________

Closing the Lesson
___ Communicating about Algebra, Student's Edition page 376
___ Guided Practice Exercises, Student's Edition page 377

Homework Assignment, pages 377–379
___ Basic/Average: Ex. 7–12, 13–19 odd, 25–33 odd, 35–36, 48–50 odd
___ Above Average: Ex. 7–12, 17–31 odd, 33–39, 44–52 odd
___ Advanced: Ex. 7–12, 22–28, 29, 31, 35–39, 50–55

Reteaching the Lesson
___ Extra Practice Copymasters page 49
___ Reteaching Copymasters page 49

Extending the Lesson
___ Cultural Diversity Extensions page 7

Notes

Lesson Plan 7.6
pages 381–386

1 day for Basic Course, 1 day for Full Course

Teacher's Name __________ *Class* ______ *Date* ______ *Room* ______

Goals
1. Solve a system of linear inequalities by graphing.
2. Model a real-life situation using a system of linear inequalities.

State/Local Objectives __________

NCTM Curriculum Standards: Problem Solving, Communication, Connections, Discrete Mathematics, Geometry

✓ **Check items you wish to use for this lesson.**

Introducing the Lesson
___ Teaching Tools: Problem of the Day copymasters page 37, or Teacher's Edition page 381
___ Teaching Tools: Warm-Up Exercises copymasters page 67, or Teacher's Edition page 381

Teaching the Lesson using the following:
___ Teaching Tools: transparencies pages 4, 11, 12

Notes for substitute teacher __________

Closing the Lesson
___ Communicating about Algebra, Student's Edition page 383
___ Guided Practice Exercises, Student's Edition page 384

Homework Assignment, pages 384–386
___ Basic/Average: Ex. 7–12, 13–27 odd, 35–38, 41–47 odd
___ Above Average: Ex. 7–12, 13–27 odd, 35–39, 43, 47–49
___ Advanced: Ex. 7–12, 13–31 odd, 33–39, 48, 49

Reteaching the Lesson
___ Extra Practice Copymasters page 50
___ Reteaching Copymasters page 50

Extending the Lesson
___ Enrichment, Teacher's Edition page 386

Notes

Lesson Plan 7.7
pages 387–392

2 days for Basic Course, 2 days for Full Course

Teacher's Name ______________________ *Class* __________ *Date* __________ *Room* __________

Goals
1. Solve a linear programming problem.
2. Model a real-life situation using linear programming.

State/Local Objectives ______________________________

NCTM Curriculum Standards: Problem Solving, Communication, Connections, Discrete Mathematics, Geometry

✓ **Check items you wish to use for this lesson.**

Introducing the Lesson
___ Teaching Tools: Problem of the Day copymasters page 37, or Teacher's Edition page 387
___ Teaching Tools: Warm-Up Exercises copymasters page 68, or Teacher's Edition page 387

Notes for substitute teacher ______________________________

Closing the Lesson
___ Communicating about Algebra, Student's Edition page 389
___ Guided Practice Exercises, Student's Edition page 390

Homework Assignment, pages 390–392
___ Basic/Average: Ex. 7–25 odd, 33, 35–41 odd
___ Above Average: Ex. 7–13 odd, 15–25 odd, 26, 35, 43
___ Advanced: Ex. 15–26, 37, 43–44

Reteaching the Lesson
___ Extra Practice Copymasters page 51
___ Reteaching Copymasters page 51

Extending the Lesson
___ Enrichment, Teacher's Edition page 392

Notes

Lesson Plan 8.1
pages 400–405

1 day for Basic Course, 1 day for Full Course

Teacher's Name ________ *Class* ________ *Date* ________ *Room* ________

Goals
1. Use the multiplication properties of exponents to evaluate powers and simplify expressions.
2. Use powers and the exponential change equations as models.

State/Local Objectives ________

NCTM Curriculum Standards: Problem Solving, Communication, Connections, Structure, Geometry

✓ **Check items you wish to use for this lesson.**

Introducing the Lesson
___ Teaching Tools: Problem of the Day copymasters page 38, or Teacher's Edition page 400
___ Teaching Tools: Warm-Up Exercises copymasters page 68, or Teacher's Edition page 400

Teaching the Lesson using the following:
___ Common-Error Alert, Teacher's Edition page 401
___ Common-Error Alert, Teacher's Edition page 404
___ Color Transparencies pages 39, 40
___ Teaching Tools: transparencies page 5

Notes for substitute teacher ________

Closing the Lesson
___ Communicating about Algebra, Student's Edition page 402
___ Extend Communicating, Teacher's Edition page 403
___ Guided Practice Exercises, Student's Edition page 403

Homework Assignment, pages 403–405
___ Basic/Average: Ex. 9–48 multiples of 3, 55–77 odd
___ Above Average: Ex. 21–77 odd, 78
___ Advanced: Ex. 23–77 odd, 78–80

Reteaching the Lesson
___ Extra Practice Copymasters page 52
___ Reteaching Copymasters page 52
___ Alternative Assessment: Math Log pages 51–53

Extending the Lesson
___ Enrichment, Teacher's Edition page 405
___ Cultural Diversity Extensions page 8

Notes

Lesson Plan 8.2
pages 406–411

2 days for Basic Course, 1 day for Full Course

Teacher's Name ______________________ *Class* __________ *Date* __________ *Room* __________

Goals

1. Use negative and zero exponents in algebraic expressions.
2. Use powers as models.

State/Local Objectives ____

____

NCTM Curriculum Standards: Problem Solving, Communication, Connections, Geometry, Structure

✓ **Check items you wish to use for this lesson.**

Introducing the Lesson
___ Teaching Tools: Problem of the Day copymasters page 38, or Teacher's Edition page 406
___ Teaching Tools: Warm-Up Exercises copymasters page 68, or Teacher's Edition page 406

Teaching the Lesson using the following:
___ Technology for Example 4, Teacher's Edition page 408
___ Common-Error Alert, Teacher's Edition page 410
___ Teaching Tools: transparencies page 14

Notes for substitute teacher ____

____

____

Closing the Lesson
___ Communicating about Algebra, Student's Edition page 408
___ Guided Practice Exercises, Student's Edition page 409

Homework Assignment, pages 409–410
___ Basic/Average: Ex. 7–45 odd, 46–47, 53–63 odd, 65–68
___ Above Average: Ex. 7–45 odd, 46–49, 53–63 odd, 65–68
___ Advanced: Ex. 19–45 odd, 46–52, 65–71

Reteaching the Lesson
___ Extra Practice Copymasters page 53
___ Reteaching Copymasters page 53

Extending the Lesson
___ Enrichment, Teacher's Edition page 411
___ Technology: Using Calculators and Computers page 59

Notes

____

____

____

Lesson Plan 8.3
pages 413–418

1 day for Basic Course, 1 day for Full Course

Teacher's Name ______________________ *Class* __________ *Date* __________ *Room* __________

Goals
1. Use the division properties of exponents to evaluate powers and simplify expressions.
2. Use powers as models.

State/Local Objectives ______________________

NCTM Curriculum Standards: Problem Solving, Communication, Structure, Connections, Reasoning

✓ **Check items you wish to use for this lesson.**

Introducing the Lesson
___ Teaching Tools: Problem of the Day copymasters page 38, or Teacher's Edition page 413
___ Teaching Tools: Warm-Up Exercises copymasters page 69, or Teacher's Edition page 413

Notes for substitute teacher ______________________

Closing the Lesson
___ Communicating about Algebra, Student's Edition page 415
___ Guided Practice Exercises, Student's Edition page 416

Homework Assignment, pages 416–417
___ Basic/Average: Ex. 5–27 odd, 31, 33, 35–47 odd, 53–56
___ Above Average: Ex. 5–29 odd, 31–34, 45–48, 53–56
___ Advanced: Ex. 15–29 odd, 30–34, 47–56

Reteaching the Lesson
___ Extra Practice Copymasters page 54
___ Reteaching Copymasters page 54

Extending the Lesson
___ Applications Handbook pages 29–31
___ Enrichment: Research, Teacher's Edition page 418

Notes

Lesson Plan 8.4
pages 419–424

1 day for Basic Course, 1 day for Full Course

Teacher's Name ______________________ *Class* __________ *Date* __________ *Room* __________

Goals
1. Use scientific notation to express large and small numbers.
2. Perform operations with numbers in scientific notation, with and without a calculator.

State/Local Objectives ______________________

NCTM Curriculum Standards: Problem Solving, Communication, Connections, Technology

✓ **Check items you wish to use for this lesson.**

Introducing the Lesson
___ Teaching Tools: Problem of the Day copymasters page 39, or Teacher's Edition page 419
___ Teaching Tools: Warm-Up Exercises copymasters page 69, or Teacher's Edition page 419

Teaching the Lesson using the following:
___ Common-Error Alert, Teacher's Edition page 420
___ Technology for Example 4, Teacher's Edition page 420
___ Alternative Approach, Teacher's Edition page 423
___ Color Transparencies pages 40, 41

Notes for substitute teacher ______________________

Closing the Lesson
___ Communicating about Algebra, Student's Edition page 421
___ Extend Communicating: Cooperative Learning, Teacher's Edition page 421
___ Guided Practice Exercises, Student's Edition page 422

Homework Assignment, pages 422–423
___ Basic/Average: Ex. 5–49 odd, 50, 53–54
___ Above Average: Ex. 5–29 odd, 33–43 odd, 45–50, 52–54
___ Advanced: Ex. 5–43 odd, 44–52, 55–56

Reteaching the Lesson
___ Extra Practice Copymasters page 55
___ Reteaching Copymasters page 55

Extending the Lesson
___ Enrichment: Research, Teacher's Edition page 424

Notes

Lesson Plan 8.5
pages 426–431

1 day for Basic Course, 1 day for Full Course

Teacher's Name ______________________ *Class* __________ *Date* __________ *Room* __________

Goal Use scientific notation to solve real-life problems.

State/Local Objectives ______________________

NCTM Curriculum Standards: Problem Solving, Communication, Connections, Reasoning

✓ **Check items you wish to use for this lesson.**

Introducing the Lesson
___ Teaching Tools: Problem of the Day copymasters page 39, or Teacher's Edition page 426
___ Teaching Tools: Warm-Up Exercises copymasters page 69, or Teacher's Edition page 426

Teaching the Lesson using the following:
___ Extension for Examples 1–3, Teacher's Edition page 426
___ Color Transparencies page 41
___ Teaching Tools: transparencies page 3

Notes for substitute teacher ______________________

Closing the Lesson
___ Communicating about Algebra, Student's Edition page 428
___ Extend Communicating, Teacher's Edition page 428
___ Guided Practice Exercises, Student's Edition page 429

Homework Assignment, pages 429–431
___ Basic/Average: Ex. 1–6, 7–10, 13, 14, 16, 17, 19–31 odd, 33–34
___ Above Average: Ex. 1–6, 7–17, 19–31 odd, 33–35
___ Advanced: Ex. 1–6, 9–17, 19–31 odd, 33–36

Reteaching the Lesson
___ Extra Practice Copymasters page 56
___ Reteaching Copymasters page 56

Extending the Lesson
___ Enrichment, Teacher's Edition page 431
___ Writing, Teacher's Edition page 431
___ Applications Handbook pages 9, 11, 12

Notes

Lesson Plan 8.6
pages 432–437

1 day for Basic Course, 2 days for Full Course

Teacher's Name ______________________ *Class* __________ *Date* __________ *Room* __________

Goals
1. Use the compound interest formula.
2. Use models for exponential growth to solve real-life problems.

State/Local Objectives ____

NCTM Curriculum Standards: Problem Solving, Communication, Connections, Geometry

✓ **Check items you wish to use for this lesson.**

Introducing the Lesson
___ Teaching Tools: Problem of the Day copymasters page 40, or Teacher's Edition page 432
___ Teaching Tools: Warm-Up Exercises copymasters page 70, or Teacher's Edition page 432

Teaching the Lesson using the following:
___ Common-Error Alert, Teacher's Edition page 436
___ Extension for Example 2, Teacher's Edition page 433
___ Teaching Tools: transparencies pages 1, 2, 3

Notes for substitute teacher ____

Closing the Lesson
___ Communicating about Algebra, Student's Edition page 434
___ Extend Communicating, Teacher's Edition page 434
___ Guided Practice Exercises, Student's Edition page 435

Homework Assignment, pages 435–436
___ Basic/Average: Ex. 5, 7, 9–14, 17, 19, 27, 29, 31–33
___ Above Average: Ex. 5–14, 16–20, 23–31 odd, 32–36
___ Advanced: Ex. 9–20, 31–36

Reteaching the Lesson
___ Extra Practice Copymasters page 57
___ Reteaching Copymasters page 57

Extending the Lesson
___ Enrichment, Teacher's Edition page 437
___ Technology: Using Calculators and Computers page 62

Notes

Lesson Plan 8.7
pages 439–444

3 days for Basic Course, 3 days for Full Course

Teacher's Name ______________________ *Class* __________ *Date* __________ *Room* __________

Goal Use models for exponential growth and decay to solve real-life problems.

State/Local Objectives ____

____

NCTM Curriculum Standards: Problem Solving, Communication, Connections, Geometry

✓ **Check items you wish to use for this lesson.**

Introducing the Lesson
___ Teaching Tools: Problem of the Day copymasters page 40, or Teacher's Edition page 439
___ Teaching Tools: Warm-Up Exercises copymasters page 70, or Teacher's Edition page 439

Notes for substitute teacher ____

____

____

Closing the Lesson
___ Communicating about Algebra, Student's Edition page 441
___ Extend Communicating, Teacher's Edition page 441
___ Guided Practice Exercises, Student's Edition page 442

Homework Assignment, pages 442–443
___ Basic/Average: Ex. 5–7, 10–17
___ Above Average: Ex. 5–9, 10–17
___ Advanced: Ex. 7–19

Reteaching the Lesson
___ Extra Practice Copymasters page 58
___ Reteaching Copymasters page 58

Extending the Lesson
___ Enrichment, Teacher's Edition page 443

Notes

____

____

____

Lesson Plan 9.1
pages 452–457

1 day for Basic Course, 1 day for Full Course

Teacher's Name ______________________ *Class* __________ *Date* __________ *Room* __________

Goals
1. Evaluate and approximate square roots.
2. Use the Pythagorean Theorem.

State/Local Objectives ______________________

____

NCTM Curriculum Standards: Problem Solving, Communication, Connections, Technology

✓ **Check items you wish to use for this lesson.**

Introducing the Lesson
___ Teaching Tools: Problem of the Day copymasters page 40, or Teacher's Edition page 452
___ Teaching Tools: Warm-Up Exercises copymasters page 70, or Teacher's Edition page 452

Teaching the Lesson using the following:
___ Extension for Example 2, Teacher's Edition page 453
___ Extension for Example 5, Teacher's Edition page 453
___ Technology for Example 4, Teacher's Edition page 453

Notes for substitute teacher ______________________

____

____

Closing the Lesson
___ Communicating about Algebra, Student's Edition page 454
___ Extend Communicating, Teacher's Edition page 455
___ Guided Practice Exercises, Student's Edition page 455

Homework Assignment, pages 455–457
___ Basic/Average: Ex. 7–37 odd, 41, 47, 49, 53, 57
___ Above Average: Ex. 7–41 odd, 43, 44, 47–59 odd
___ Advanced: Ex. 13–41 odd, 43–50, 61–64

Reteaching the Lesson
___ Extra Practice Copymasters page 59
___ Reteaching Copymasters page 59
___ Alternative Assessment: Math Log page 9

Extending the Lesson
___ Enrichment, Teacher's Edition page 457

Notes

____

____

____

Lesson Plan 9.2
pages 458–464

1 day for Basic Course, 1 day for Full Course

Teacher's Name ______________________ *Class* __________ *Date* __________ *Room* __________

Goals
1. Solve a quadratic equation by finding square roots.
2. Use quadratic models in real-life settings.

State/Local Objectives ____

____

NCTM Curriculum Standards: Problem Solving, Communication, Connections, Geometry

✓ **Check items you wish to use for this lesson.**

Introducing the Lesson
___ Teaching Tools: Problem of the Day copymasters page 40, or Teacher's Edition page 458
___ Teaching Tools: Warm-Up Exercises copymasters page 71, or Teacher's Edition page 458

Teaching the Lesson using the following:
___ Common-Error Alert, Teacher's Edition page 458
___ Technology for Example 3, Teacher's Edition page 459
___ Color Transparencies pages 43, 44
___ Teaching Tools: transparencies page 4

Notes for substitute teacher ____

____

____

Closing the Lesson
___ Communicating about Algebra, Student's Edition page 460
___ Extend Communicating, Teacher's Edition pages 460–461
___ Guided Practice Exercises, Student's Edition page 461

Homework Assignment, pages 461–463
___ Basic/Average: Ex. 9–25 odd, 33, 35–36, 39–40, 41–51 odd
___ Above Average: Ex. 9–33 odd, 34–38, 39–49 odd, 51–52
___ Advanced: Ex. 17–33 odd, 34–38, 39–49 odd, 51–54

Reteaching the Lesson
___ Extra Practice Copymasters page 60
___ Reteaching Copymasters page 60

Extending the Lesson
___ Enrichment, Teacher's Edition page 463
___ Applications Handbook pages 44, 45

Notes

____

____

____

Lesson Plan 9.3
pages 465–470

1 day for Basic Course, 1 day for Full Course

Teacher's Name ______________________ *Class* __________ *Date* __________ *Room* __________

Goals
1. Sketch the graph of a quadratic equation.
2. Use quadratic models.

State/Local Objectives ____

NCTM Curriculum Standards: Problem Solving, Communication, Connections, Geometry

✓ **Check items you wish to use for this lesson.**

Introducing the Lesson
___ Teaching Tools: Problem of the Day copymasters page 41, or Teacher's Edition page 465
___ Teaching Tools: Warm-Up Exercises copymasters page 71, or Teacher's Edition page 465

Teaching the Lesson using the following:
___ Common-Error Alert, Teacher's Edition page 465
___ Color Transparencies pages 45, 46
___ Teaching Tools: transparencies pages 1, 2, 4, 13

Notes for substitute teacher ____

Closing the Lesson
___ Communicating about Algebra, Student's Edition page 467
___ Guided Practice Exercises, Student's Edition page 468

Homework Assignment, pages 468–469
___ Basic/Average: Ex. 7–33 odd, 41–45, 51–56
___ Above Average: Ex. 11–29 odd, 39–45, 51–56
___ Advanced: Ex. 11–29 odd, 37–48, 51–56, 61, 63–65

Reteaching the Lesson
___ Extra Practice Copymasters page 61
___ Reteaching Copymasters page 61

Extending the Lesson
___ Applications Handbook pages 46–48
___ Technology: Using Calculators and Computers page 65

Notes

Lesson Plan 9.4
pages 472–477

1 day for Basic Course, 1 day for Full Course

Teacher's Name ________________ *Class* ________ *Date* ________ *Room* ________

Goals
1. Use the quadratic formula to solve a quadratic equation.
2. Use quadratic models.

State/Local Objectives ________________________

NCTM Curriculum Standards: Problem Solving, Communication, Connections

✓ **Check items you wish to use for this lesson.**

Introducing the Lesson
- ___ Teaching Tools: Problem of the Day copymasters page 41, or Teacher's Edition page 472
- ___ Teaching Tools: Warm-Up Exercises copymasters page 71, or Teacher's Edition page 472

Teaching the Lesson using the following:
- ___ Common-Error Alert, Teacher's Edition page 473
- ___ Color Transparencies page 46
- ___ Teaching Tools: transparencies pages 4, 13

Notes for substitute teacher ________________________

Closing the Lesson
- ___ Communicating about Algebra, Student's Edition page 474
- ___ Extend Communicating, Teacher's Edition page 475
- ___ Guided Practice Exercises, Student's Edition page 475

Homework Assignment, pages 475–477
- ___ Basic/Average: Ex. 7–31 odd, 39, 41–42, 43–48
- ___ Above Average: Ex. 7–31 odd, 34, 36, 39–49 odd
- ___ Advanced: Ex. 15–33 odd, 34–36, 39–40, 43–50

Reteaching the Lesson
- ___ Extra Practice Copymasters page 62
- ___ Reteaching Copymasters page 62

Extending the Lesson
- ___ Enrichment, Teacher's Edition page 477
- ___ Applications Handbook pages 46–48
- ___ Technology: Using Calculators and Computers page 68

Notes

Lesson Plan 9.5
pages 479–485

2 days for Basic Course, 2 days for Full Course

Teacher's Name ______________________ *Class* __________ *Date* __________ *Room* __________

Goals
1. Find the number of solutions of a quadratic equation by using the discriminant.
2. Use the discriminant in real-life models.

State/Local Objectives ____

NCTM Curriculum Standards: Problem Solving, Communication, Connections

✓ **Check items you wish to use for this lesson.**

Introducing the Lesson
___ Teaching Tools: Problem of the Day copymasters page 41, or Teacher's Edition page 479
___ Teaching Tools: Warm-Up Exercises copymasters page 72, or Teacher's Edition page 479

Teaching the Lesson using the following:
___ Color Transparencies pages 47, 48, 49
___ Teaching Tools: transparencies pages 3, 13

Notes for substitute teacher ____

Closing the Lesson
___ Communicating about Algebra, Student's Edition page 481
___ Guided Practice Exercises, Student's Edition page 482

Homework Assignment, pages 482–484
___ Basic/Average: Ex. 7–12, 15, 17, 18, 21–22, 23–25, 27–30
___ Above Average: Ex. 7–15, 17, 18, 21–22, 23–25, 27–34
___ Advanced: Ex. 11, 13–18, 20–34

Reteaching the Lesson
___ Extra Practice Copymasters page 63
___ Reteaching Copymasters page 63

Extending the Lesson
___ Enrichment, Teacher's Edition page 484
___ Applications Handbook pages 46–48
___ Technology: Using Calculators and Computers page 70

Notes

Lesson Plan 9.6

pages 486–491

1 day for Basic Course, 1 day for Full Course

Teacher's Name ________________ *Class* ________ *Date* ________ *Room* ________

Goals

1. Sketch the graph of a quadratic inequality.
2. Use quadratic inequalities as real-life models.

State/Local Objectives ________________

NCTM Curriculum Standards: Problem Solving, Communication, Connections, Geometry

✓ **Check items you wish to use for this lesson.**

Introducing the Lesson
___ Teaching Tools: Problem of the Day copymasters page 42, or Teacher's Edition page 486
___ Teaching Tools: Warm-Up Exercises copymasters page 72, or Teacher's Edition page 486

Teaching the Lesson using the following:
___ Color Transparencies page 50
___ Teaching Tools: transparencies pages 4, 11, 12, 13

Notes for substitute teacher ________________

Closing the Lesson
___ Communicating about Algebra, Student's Edition page 488
___ Guided Practice Exercises, Student's Edition page 489

Homework Assignment, pages 489–491
___ Basic/Average: Ex. 5–7, 9–14, 15–23, 29–30, 31–41
___ Above Average: Ex. 5–7, 9–14, 15–23 odd, 28–30, 31–43 odd
___ Advanced: Ex. 9–17, 22–27, 29–30, 31–43 odd, 44

Reteaching the Lesson
___ Extra Practice Copymasters page 64
___ Reteaching Copymasters page 64

Extending the Lesson
___ Enrichment, Teacher's Edition page 491

Notes

Lesson Plan 9.7
pages 492–497

2 days for Basic Course, 2 days for Full Course

Teacher's Name ______ *Class* ______ *Date* ______ *Room* ______

Goals
1. Choose a model that best fits a collection of data.
2. Use models in real-life settings.

State/Local Objectives ______

NCTM Curriculum Standards: Problem Solving, Communication, Connections, Geometry

✓ **Check items you wish to use for this lesson.**

Introducing the Lesson
___ Teaching Tools: Problem of the Day copymasters page 42, or Teacher's Edition page 492
___ Teaching Tools: Warm-Up Exercises copymasters page 72, or Teacher's Edition page 492

Teaching the Lesson using the following:
___ Color Transparencies pages 51, 52
___ Teaching Tools: transparencies pages 4, 10, 13, 14

Notes for substitute teacher ______

Closing the Lesson
___ Communicating about Algebra, Student's Edition page 494
___ Guided Practice Exercises, Student's Edition page 495

Homework Assignment, pages 495–497
___ Basic/Average: Ex. 5–14, 19–29 odd
___ Above Average: Ex. 5–8, 9–21 odd, 23–29 odd, 31
___ Advanced: Ex. 5–8, 15–18, 19–22, 23–29 odd, 31

Reteaching the Lesson
___ Extra Practice Copymasters page 65
___ Reteaching Copymasters page 65

Extending the Lesson
___ Applications Handbook pages 49–51
___ Technology: Using Calculators and Computers page 76

Notes

Lesson Plan 10.1
pages 508–514

1 day for Basic Course, 1 day for Full Course

Teacher's Name ______________________ *Class* __________ *Date* __________ *Room* __________

Goals
1. Add and subtract polynomials.
2. Use polynomials as models.

State/Local Objectives ______________________

NCTM Curriculum Standards: Problem Solving, Communication, Connections, Geometry

✓ **Check items you wish to use for this lesson.**

Introducing the Lesson
___ Teaching Tools: Problem of the Day copymasters page 42, or Teacher's Edition page 508
___ Teaching Tools: Warm-Up Exercises copymasters page 73, or Teacher's Edition page 508

Teaching the Lesson using the following:
___ Alternative Approach for Examples 1–2, Teacher's Edition page 509
___ Common-Error Alert, Teacher's Edition page 512
___ Color Transparencies pages 55, 56, 57

Notes for substitute teacher ______________________

Closing the Lesson
___ Communicating about Algebra, Student's Edition page 511
___ Extend Communicating: Writing, Teacher's Edition page 511
___ Guided Practice Exercises, Student's Edition page 512

Homework Assignment, pages 512–514
___ Basic/Average: Ex. 7–31 odd, 35–38, 41–49 odd, 51, 52
___ Above Average: Ex. 7-33 odd, 35–40, 41–49 odd, 51–53
___ Advanced: Ex. 7–39 odd, 40, 41–49 odd, 51–53

Reteaching the Lesson
___ Extra Practice Copymasters page 66
___ Reteaching Copymasters page 66
___ Alternative Assessment: Math Log pages 58–61

Extending the Lesson
___ Enrichment, Teacher's Edition page 514

Notes

Lesson Plan 10.2
pages 515–520

2 days for Basic Course, 1 day for Full Course

Teacher's Name ______________________ *Class* __________ *Date* __________ *Room* __________

Goals
1. Multiply two polynomials using the Distributive Properties and the FOIL Method.
2. Use polynomial multiplication in real-life settings.

State/Local Objectives ______________________________

NCTM Curriculum Standards: Problem Solving, Communication, Connections, Geometry

✓ **Check items you wish to use for this lesson.**

Introducing the Lesson
___ Teaching Tools: Problem of the Day copymasters page 43, or Teacher's Edition page 515
___ Teaching Tools: Warm-Up Exercises copymasters page 73, or Teacher's Edition page 515

Teaching the Lesson using the following:
___ Color Transparencies page 58
___ Teaching Tools: transparencies page 16

Notes for substitute teacher ______________________________

Closing the Lesson
___ Communicating about Algebra, Student's Edition page 517
___ Extend Communicating, Teacher's Edition pages 517–518
___ Guided Practice Exercises, Student's Edition page 518

Homework Assignment, pages 518–520
___ Basic/Average: Ex. 1–4, 7–31 odd, 43–45, 47–61 odd
___ Above Average: Ex. 1–4, 9–35 odd, 38–43, 46, 48–63 multiples of 3, 64
___ Advanced: Ex. 1–4, 11–37 odd, 38–46, 63–65

Reteaching the Lesson
___ Extra Practice Copymasters page 67
___ Reteaching Copymasters page 67

Extending the Lesson
___ Enrichment, Teacher's Edition pages 520–521

Notes

Lesson Plan 10.3
pages 522–527

3 days for Basic Course, 3 days for Full Course

Teacher's Name ______________________ *Class* __________ *Date* __________ *Room* __________

Goals
1. Use patterns for the product of a sum and difference and for the square of a binomial.
2. Use special-product patterns in real-life models.

State/Local Objectives ______________________

NCTM Curriculum Standards: Problem Solving, Communication, Connections, Geometry

✓ **Check items you wish to use for this lesson.**

Introducing the Lesson
___ Teaching Tools: Problem of the Day copymasters page 43, or Teacher's Edition page 522
___ Teaching Tools: Warm-Up Exercises copymasters page 73, or Teacher's Edition page 522

Teaching the Lesson using the following:
___ Teaching Tools: transparencies page 16

Notes for substitute teacher ______________________

Closing the Lesson
___ Communicating about Algebra, Student's Edition page 524
___ Extend Communicating, Teacher's Edition page 525
___ Guided Practice Exercises, Student's Edition page 525

Homework Assignment, pages 525–527
___ Basic/Average: Ex. 5–27 odd, 31, 33, 36–44, 45–53 odd, 56
___ Above Average: Ex. 5–27 odd, 29–34, 45–55 odd, 56
___ Advanced: Ex. 9–27 odd, 29–38, 45–56

Reteaching the Lesson
___ Extra Practice Copymasters page 68
___ Reteaching Copymasters page 68

Extending the Lesson
___ Enrichment, Teacher's Edition page 527
___ Applications Handbook pages 13, 14

Notes

Lesson Plan 10.4
pages 528–533

3 days for Basic Course, 3 days for Full Course

Teacher's Name ____________________ *Class* __________ *Date* __________ *Room* __________

Goals
1. Factor polynomials, including the difference of two squares and perfect square trinomials.
2. Use factoring in real-life models.

State/Local Objectives ____

NCTM Curriculum Standards: Problem Solving, Communication, Connections

✓ **Check items you wish to use for this lesson.**

Introducing the Lesson
___ Teaching Tools: Problem of the Day copymasters page 43, or Teacher's Edition page 528
___ Teaching Tools: Warm-Up Exercises copymasters page 74, or Teacher's Edition page 528

Teaching the Lesson using the following:
___ Color Transparencies page 58

Notes for substitute teacher ____

Closing the Lesson
___ Communicating about Algebra, Student's Edition page 530
___ Guided Practice Exercises, Student's Edition page 531

Homework Assignment, pages 531–533
___ Basic/Average: Ex. 5–27 odd, 29–33, 37–49 odd, 51
___ Above Average: Ex. 5–27 odd, 29–34, 36–38, 39–53 odd
___ Advanced: Ex. 11–27 odd, 29–37, 45–51 odd, 52, 53

Reteaching the Lesson
___ Extra Practice Copymasters page 69
___ Reteaching Copymasters page 69

Extending the Lesson
___ Enrichment, Teacher's Edition page 532

Notes

Lesson Plan 10.5
pages 535–540

2 days for Basic Course, 2 days for Full Course

Teacher's Name ______________________________ *Class* ___________ *Date* ________ *Room* ________

Goals
1. Factor a quadratic trinomial or recognize that it cannot be factored.
2. Use factoring in real-life models.

State/Local Objectives ______________________________

NCTM Curriculum Standards: Problem Solving, Communication, Connections, Structure, Geometry

✓ **Check items you wish to use for this lesson.**

Introducing the Lesson
___ Teaching Tools: Problem of the Day copymasters page 44, or Teacher's Edition page 535
___ Teaching Tools: Warm-Up Exercises copymasters page 74, or Teacher's Edition page 535

Teaching the Lesson using the following:
___ Color Transparencies page 59

Notes for substitute teacher ______________________________

Closing the Lesson
___ Communicating about Algebra, Student's Edition page 537
___ Guided Practice Exercises, Student's Edition page 538

Homework Assignment, pages 538–540
___ Basic/Average: Ex. 5–7, 8–10, 11–19 odd, 21–23, 37–46, 47–48, 51–55 odd, 56
___ Above Average: Ex. 5–7, 9, 11–19 odd, 21–33, 35, 37–46, 47, 49–57
___ Advanced: Ex. 5, 7, 9, 11–19 odd, 21–33, 35, 37–46, 47–48, 49–50, 56, 57–60

Reteaching the Lesson
___ Extra Practice Copymasters page 70
___ Reteaching Copymasters page 70

Extending the Lesson
___ Enrichment, Teacher's Edition pages 539–540

Notes

Lesson Plan 10.6
pages 541–547

1 day for Basic Course, 1 day for Full Course

Teacher's Name ______________________ *Class* __________ *Date* __________ *Room* __________

Goals 1. Use factoring to solve a quadratic equation.
2. Use factoring with real-life models.

State/Local Objectives ______________________________

NCTM Curriculum Standards: Problem Solving, Communication, Connections, Technology, Reasoning

✓ **Check items you wish to use for this lesson.**

Introducing the Lesson
___ Teaching Tools: Problem of the Day copymasters page 44, or Teacher's Edition page 541
___ Teaching Tools: Warm-Up Exercises copymasters page 74, or Teacher's Edition page 541

Teaching the Lesson using the following:
___ Technology for Examples 1–3, Teacher's Edition page 542
___ Common-Error Alert, Teacher's Edition page 543
___ Teaching Tools: transparencies pages 1, 2

Notes for substitute teacher ______________________________

Closing the Lesson
___ Communicating about Algebra, Student's Edition page 543
___ Guided Practice Exercises, Student's Edition page 544

Homework Assignment, pages 544–547
___ Basic/Average: Ex. 7–15 odd, 17–24, 25–33 odd, 38, 39, 41–51 odd, 59–65 odd
___ Above Average: Ex. 11–15 odd, 17–24, 25–37 odd, 38–39, 42–63 multiples of 3, 66–68
___ Advanced: Ex. 17–24, 25–35 odd, 36–39, 41–61 odd, 62–68

Reteaching the Lesson
___ Extra Practice Copymasters page 71
___ Reteaching Copymasters page 71

Extending the Lesson
___ Cultural Diversity Extensions page 10
___ Technology: Using Calculators and Computers page 78

Notes

Lesson Plan 10.7
pages 549–554

1 day for Basic Course, 1 day for Full Course

Teacher's Name ______________________ *Class* __________ *Date* __________ *Room* __________

Goal Solve quadratic equations by completing the square.

State/Local Objectives ______________________

NCTM Curriculum Standards: Problem Solving, Communication, Connections, Structure, Reasoning

✓ **Check items you wish to use for this lesson.**

Introducing the Lesson
___ Teaching Tools: Problem of the Day copymasters page 44, or Teacher's Edition page 549
___ Teaching Tools: Warm-Up Exercises copymasters page 75, or Teacher's Edition page 549

Teaching the Lesson using the following:
___ Technology for Example 1, Teacher's Edition page 550
___ Teaching Tools: transparencies pages 11, 12, 13

Notes for substitute teacher ______________________

Closing the Lesson
___ Communicating about Algebra, Student's Edition page 551
___ Guided Practice Exercises, Student's Edition page 552

Homework Assignment, pages 552–553
___ Basic/Average: Ex. 5–43 odd, 44, 51–63 multiples of 3, 67–70
___ Above Average: Ex. 17–27 odd, 35–43 odd, 44–48, 49–65 odd, 71, 73, 75–76
___ Advanced: Ex. 17–27 odd, 35–43 odd, 44–48, 54, 60, 66, 71–77

Reteaching the Lesson
___ Extra Practice Copymasters page 72
___ Reteaching Copymasters page 72

Extending the Lesson
___ Enrichment, Teacher's Edition page 554

Notes

Lesson Plan 11.1
pages 562–567

1 day for Basic Course, 1 day for Full Course

Teacher's Name ______________________ *Class* __________ *Date* ________ *Room* ________

Goals
1. Solve proportions.
2. Use proportions to solve real-life problems.

State/Local Objectives ______________________

NCTM Curriculum Standards: Problem Solving, Communication, Connections

✓ **Check items you wish to use for this lesson.**

Introducing the Lesson
___ Teaching Tools: Problem of the Day copymasters page 45, or Teacher's Edition page 562
___ Teaching Tools: Warm-Up Exercises copymasters page 75, or Teacher's Edition page 562

Teaching the Lesson using the following:
___ Common-Error Alert, Teacher's Edition page 563
___ Extension for Example 4, Teacher's Edition page 563
___ Color Transparencies page 69
___ Teaching Tools: transparencies page 3

Notes for substitute teacher ______________________

Closing the Lesson
___ Communicating about Algebra, Student's Edition page 564
___ Guided Practice Exercises, Student's Edition page 565

Homework Assignment, pages 565–566
___ Basic/Average: Ex. 7–25 odd, 29, 30, 31–42, 49–50
___ Above Average: Ex. 7–27 odd, 29–50
___ Advanced: Ex. 7–27 odd, 28–30, 47–50

Reteaching the Lesson
___ Extra Practice Copymasters page 73
___ Reteaching Copymasters page 73
___ Alternative Assessment: Math Log pages 62–64

Extending the Lesson
___ Enrichment, Teacher's Edition page 567
___ Cultural Diversity Extensions page 11

Notes

Lesson Plan 11.2
pages 568–573

1 day for Basic Course, 1 day for Full Course

Teacher's Name ____________________ *Class* ________ *Date* ________ *Room* ________

Goals
1. Solve percent problems.
2. Use percents in real-life problems.

State/Local Objectives ____

____

NCTM Curriculum Standards: Problem Solving, Communication, Reasoning, Connections

✓ **Check items you wish to use for this lesson.**

Introducing the Lesson
___ Teaching Tools: Problem of the Day copymasters page 45, or Teacher's Edition page 568
___ Teaching Tools: Warm-Up Exercises copymasters page 76, or Teacher's Edition page 568

Teaching the Lesson using the following:
___ Alternative Approach, Teacher's Edition page 569
___ Color Transparencies page 61
___ Teaching Tools: transparencies page 3

Notes for substitute teacher ____

____

____

Closing the Lesson
___ Communicating about Algebra, Student's Edition page 570
___ Guided Practice Exercises, Student's Edition page 571

Homework Assignment, pages 571–573
___ Basic/Average: Ex. 5–25 odd, 27–28, 31–33, 37–53 odd
___ Above Average: Ex. 5–21 odd, 23–24, 43–53 odd
___ Advanced: Ex. 11–25 odd, 27–36, 53–56

Reteaching the Lesson
___ Extra Practice Copymasters page 74
___ Reteaching Copymasters page 74

Extending the Lesson
___ Enrichment, Teacher's Edition page 573

Notes

____

____

____

Lesson Plan 11.3
pages 575–580

2 days for Basic Course, 2 days for Full Course

Teacher's Name ______________ *Class* ________ *Date* ________ *Room* ________

Goals
1. Use direct and inverse variation.
2. Use direct and inverse variation in real-life settings.

State/Local Objectives ______________________________

NCTM Curriculum Standards: Problem Solving, Communication, Connections

✓ **Check items you wish to use for this lesson.**

Introducing the Lesson
___ Teaching Tools: Problem of the Day copymasters page 46, or Teacher's Edition page 575
___ Teaching Tools: Warm-Up Exercises copymasters page 76, or Teacher's Edition page 575

Teaching the Lesson using the following:
___ Color Transparencies pages 62, 63

Notes for substitute teacher ______________________________

Closing the Lesson
___ Communicating about Algebra, Student's Edition page 577
___ Guided Practice Exercises, Student's Edition page 578

Homework Assignment, pages 578–580
___ Basic/Average: Ex. 11–27 odd, 28, 30, 32, 41–42
___ Above Average: Ex. 11–29 odd, 30–32, 41–42
___ Advanced: Ex. 11–29 odd, 30–32, 43

Reteaching the Lesson
___ Extra Practice Copymasters page 75
___ Reteaching Copymasters page 75

Extending the Lesson
___ Enrichment, Teacher's Edition page 580

Notes

Lesson Plan 11.4
pages 581–586

1 day for Basic Course, 1 day for Full Course

Teacher's Name __________ *Class* ______ *Date* ______ *Room* ______

Goals
1. Find the probability of an event.
2. Use probability in real-life settings.

State/Local Objectives __________

NCTM Curriculum Standards: Problem Solving, Communication, Connections, Discrete Mathematics, Geometry

✓ **Check items you wish to use for this lesson.**

Introducing the Lesson
___ Teaching Tools: Problem of the Day copymasters page 46, or Teacher's Edition page 581
___ Teaching Tools: Warm-Up Exercises copymasters page 77, or Teacher's Edition page 581

Teaching the Lesson using the following:
___ Color Transparencies pages 63, 64

Notes for substitute teacher __________

Closing the Lesson
___ Communicating about Algebra, Student's Edition page 583
___ Guided Practice Exercises, Student's Edition page 584

Homework Assignment, pages 584–586
___ Basic/Average: Ex. 7–11, 13–14, 17–19, 23–25
___ Above Average: Ex. 7–14, 17–22, 23–25
___ Advanced: Ex. 11–30

Reteaching the Lesson
___ Extra Practice Copymasters page 76
___ Reteaching Copymasters page 76

Extending the Lesson
___ Enrichment, Teacher's Edition pages 586–587
___ Technology: Using Calculators and Computers pages 81, 83

Notes

Lesson Plan 11.5
pages 588–593

3 days for Basic Course, 3 days for Full Course

Teacher's Name ______________________ *Class* __________ *Date* __________ *Room* __________

Goals
1. Simplify a rational expression.
2. Use rational expressions as real-life models.

State/Local Objectives ____

NCTM Curriculum Standards: Problem Solving, Communication, Structure, Connections

✓ **Check items you wish to use for this lesson.**

Introducing the Lesson
___ Teaching Tools: Problem of the Day copymasters page 46, or Teacher's Edition page 588
___ Teaching Tools: Warm-Up Exercises copymasters page 78, or Teacher's Edition page 588

Teaching the Lesson using the following:
___ Common-Error Alert, Teacher's Edition page 589
___ Color Transparencies page 64

Notes for substitute teacher ____

Closing the Lesson
___ Communicating about Algebra, Student's Edition page 590
___ Extend Communicating, Teacher's Edition page 591
___ Guided Practice Exercises, Student's Edition page 591

Homework Assignment, pages 591–593
___ Basic/Average: Ex. 5–23 odd, 29, 30, 33–36, 37–47 odd
___ Above Average: Ex. 7–27 odd, 29–36, 41–49 odd
___ Advanced: Ex. 11–27 odd, 29–37, 41–50

Reteaching the Lesson
___ Extra Practice Copymasters page 77
___ Reteaching Copymasters page 77

Extending the Lesson
___ Enrichment, Teacher's Edition pages 593

Notes

Lesson Plan 11.6
pages 594–599

3 days for Basic Course, 3 days for Full Course

Teacher's Name ______________________ *Class* __________ *Date* __________ *Room* __________

Goals
1. Multiply and divide rational expressions.
2. Use rational expressions in real-life settings.

State/Local Objectives ______________________________

NCTM Curriculum Standards: Problem Solving, Communication, Reasoning, Connections

✓ **Check items you wish to use for this lesson.**

Introducing the Lesson
___ Teaching Tools: Problem of the Day copymasters page 46, or Teacher's Edition page 594
___ Teaching Tools: Warm-Up Exercises copymasters page 78, or Teacher's Edition page 594

Teaching the Lesson using the following:
___ Extension for Example 6, Teacher's Edition page 595
___ Color Transparencies page 65

Notes for substitute teacher ______________________________

Closing the Lesson
___ Communicating about Algebra, Student's Edition page 596
___ Guided Practice Exercises, Student's Edition page 597

Homework Assignment, pages 597–599
___ Basic/Average: Ex. 7–25 odd, 31–34, 45–52, 55–56
___ Above Average: Ex. 7–27 odd, 31–36, 43–55 odd, 57–60
___ Advanced: Ex. 11–29 odd, 31–36, 49–57 odd, 59–63

Reteaching the Lesson
___ Extra Practice Copymasters page 78
___ Reteaching Copymasters page 78

Extending the Lesson
___ Enrichment, Teacher's Edition pages 599

Notes

Lesson Plan 11.7
pages 600–605

0 days for Basic Course, 1 day for Full Course

Teacher's Name ______________________ *Class* __________ *Date* __________ *Room* __________

Goals
1. Divide a polynomial by a monomial or a binomial.
2. Use polynomial division in real-life models.

State/Local Objectives ____

NCTM Curriculum Standards: Problem Solving, Communication, Connections

✓ **Check items you wish to use for this lesson.**

Introducing the Lesson
___ Teaching Tools: Problem of the Day copymasters page 47, or Teacher's Edition page 600
___ Teaching Tools: Warm-Up Exercises copymasters page 79, or Teacher's Edition page 600

Teaching the Lesson using the following:
___ Color Transparencies page 65
___ Teaching Tools: transparencies pages 1, 2

Notes for substitute teacher ____

Closing the Lesson
___ Communicating about Algebra, Student's Edition page 602
___ Extend Communicating, Teacher's Edition page 603
___ Guided Practice Exercises, Student's Edition page 603

Homework Assignment, pages 603–605
___ Basic/Average: Ex. 7–15 odd, 19–27 multiples of 3, 28, 31–41 odd, 43, 44
___ Above Average: Ex. 7–25 odd, 29–30, 31–45 odd
___ Advanced: Ex. 15–25 odd, 29–30, 31–43 odd, 45–47

Reteaching the Lesson
___ Extra Practice Copymasters page 79
___ Reteaching Copymasters page 79

Extending the Lesson
___ Enrichment, Teacher's Edition pages 605

Notes

Lesson Plan 11.8
pages 607–612

Omit for Basic Course, 1 day for Full Course

Teacher's Name ______________________ *Class* ________ *Date* ________ *Room* ________

Goals
1. Solve rational equations.
2. Use rational expressions in real-life settings.

State/Local Objectives ______________________

NCTM Curriculum Standards: Problem Solving, Communication, Structure, Connections

✓ **Check items you wish to use for this lesson.**

Introducing the Lesson
___ Teaching Tools: Problem of the Day copymasters page 47, or Teacher's Edition page 607
___ Teaching Tools: Warm-Up Exercises copymasters page 79, or Teacher's Edition page 607

Teaching the Lesson using the following:
___ Common-Error Alert, Teacher's Edition page 607
___ Color Transparencies page 66

Notes for substitute teacher ______________________

Closing the Lesson
___ Communicating about Algebra, Student's Edition page 609
___ Extend Communicating, Teacher's Edition page 609
___ Guided Practice Exercises, Student's Edition page 610

Homework Assignment, pages 610–612
___ Basic/Average: Ex. 9–23 odd, 31–47 odd, 51–59 odd
___ Above Average: Ex. 7–31 odd, 37–45 odd, 47–51, 57–63 odd
___ Advanced: Ex. 15–45 odd, 46–51, 55–65 odd

Reteaching the Lesson
___ Extra Practice Copymasters page 80
___ Reteaching Copymasters page 80

Extending the Lesson
___ Enrichment, Teacher's Edition pages 612

Notes

Lesson Plan 12.1
pages 620–625

1 day for Basic Course, 1 day for Full Course

Teacher's Name ______________________ *Class* __________ *Date* ________ *Room* ________

Goals
1. Identify functions and use function notation.
2. Identify real-life relations that are functions.

State/Local Objectives ____

NCTM Curriculum Standards: Problem Solving, Communication, Connections, Functions

✓ **Check items you wish to use for this lesson.**

Introducing the Lesson
___ Teaching Tools: Problem of the Day copymasters page 47, or Teacher's Edition page 620
___ Teaching Tools: Warm-Up Exercises copymasters page 79, or Teacher's Edition page 620

Teaching the Lesson using the following:
___ Color Transparencies pages 68, 69
___ Teaching Tools: transparencies pages 10, 13

Notes for substitute teacher ____

Closing the Lesson
___ Communicating about Algebra, Student's Edition page 622
___ Guided Practice Exercises, Student's Edition page 623

Homework Assignment, pages 623–625
___ Basic/Average: Ex. 5–8, 9–27 odd, 31–34, 43–46
___ Above Average: Ex. 5–8, 9–33 odd, 34–48
___ Advanced: Ex. 5–8, 9–31 odd, 32–48

Reteaching the Lesson
___ Extra Practice Copymasters page 81
___ Reteaching Copymasters page 81
___ Alternative Assessment: Math Log pages 65–67

Extending the Lesson
___ Enrichment, Teacher's Edition page 625

Notes

Lesson Plan 12.2
pages 626–631

2 days for Basic Course, 1 day for Full Course

Teacher's Name ______ *Class* ______ *Date* ______ *Room* ______

Goals
1. Use linear functions to solve problems.
2. Use linear functions to answer questions about real-life situations.

State/Local Objectives ______

NCTM Curriculum Standards: Problem Solving, Communication, Connections, Functions

✓ **Check items you wish to use for this lesson.**

Introducing the Lesson
___ Teaching Tools: Problem of the Day copymasters page 48, or Teacher's Edition page 626
___ Teaching Tools: Warm-Up Exercises copymasters page 80, or Teacher's Edition page 626

Teaching the Lesson using the following:
___ Color Transparencies page 70
___ Teaching Tools: transparencies page 4

Notes for substitute teacher ______

Closing the Lesson
___ Communicating about Algebra, Student's Edition page 628
___ Guided Practice Exercises, Student's Edition page 629

Homework Assignment, pages 629–631
___ Basic/Average: Ex. 5–8, 9–21 odd, 27–33 odd, 36–42, 51, 53
___ Above Average: Ex. 5–8, 11–23 odd, 27–35 odd, 36–42, 48–55 odd
___ Advanced: Ex. 5–8, 13–33 odd, 34–42, 55

Reteaching the Lesson
___ Extra Practice Copymasters page 82
___ Reteaching Copymasters page 82

Extending the Lesson
___ Enrichment, Teacher's Edition pages 631
___ Applications Handbook page 54
___ Cultural Diversity Extensions page 12

Notes

Lesson Plan 12.3
pages 633–638

1 day for Basic Course, 1 day for Full Course

Teacher's Name ______________________ *Class* __________ *Date* ________ *Room* ________

Goals
1. Transform the graph of a function.
2. Use exponential functions as models of real-life problems.

State/Local Objectives ______________________________

NCTM Curriculum Standards: Problem Solving, Communication, Connections, Functions, Geometry

✓ **Check items you wish to use for this lesson.**

Introducing the Lesson
___ Teaching Tools: Problem of the Day copymasters page 48, or Teacher's Edition page 633
___ Teaching Tools: Warm-Up Exercises copymasters page 80, or Teacher's Edition page 633

Teaching the Lesson using the following:
___ Color Transparencies pages 70, 71
___ Teaching Tools: transparencies pages 4, 14

Notes for substitute teacher ______________________________

Closing the Lesson
___ Communicating about Algebra, Student's Edition page 635
___ Extend Communicating, Teacher's Edition page 635
___ Guided Practice Exercises, Student's Edition page 636

Homework Assignment, pages 636–638
___ Basic/Average: Ex. 7–35 odd, 41, 43–46, 51–56
___ Above Average: Ex. 11–35 odd, 40–48, 53–57
___ Advanced: Ex. 13–35, 37–46, 53–57

Reteaching the Lesson
___ Extra Practice Copymasters page 83
___ Reteaching Copymasters page 83

Extending the Lesson
___ Enrichment, Teacher's Edition pages 638
___ Applications Handbook, pages 26–28, 58, 59

Notes

Lesson Plan 12.4
pages 639–644

1 day for Basic Course, 1 day for Full Course

Teacher's Name ______ *Class* ______ *Date* ______ *Room* ______

Goals
1. Sketch the graph of a quadratic function.
2. Use quadratic functions as models of real-life problems.

State/Local Objectives ______

NCTM Curriculum Standards: Problem Solving, Communication, Reasoning, Connections, Functions, Geometry

✓ **Check items you wish to use for this lesson.**

Introducing the Lesson
___ Teaching Tools: Problem of the Day copymasters page 48, or Teacher's Edition page 639
___ Teaching Tools: Warm-Up Exercises copymasters page 80, or Teacher's Edition page 639

Teaching the Lesson using the following:
___ Common-Error Alert, Teacher's Edition page 640
___ Teaching Tools: transparencies pages 4, 13

Notes for substitute teacher ______

Closing the Lesson
___ Communicating about Algebra, Student's Edition page 641
___ Guided Practice Exercises, Student's Edition page 642

Homework Assignment, pages 642–644
___ Basic/Average: Ex. 5–9 odd, 11–14, 15–33 odd, 39–40, 49–57 odd, 61–62
___ Above Average: Ex. 7, 9, 11–14, 15–35 odd, 37–42, 49–57 odd, 61–63
___ Advanced: Ex. 9, 11–14, 15–35 odd, 37–42, 49–57 odd, 61–63

Reteaching the Lesson
___ Extra Practice Copymasters page 84
___ Reteaching Copymasters page 84

Extending the Lesson
___ Enrichment, Teacher's Edition pages 644
___ Technology: Using Calculators and Computers page 84

Notes

Lesson Plan 12.5
pages 646–651

1 day for Basic Course, 1 day for Full Course

Teacher's Name ________________ *Class* ________ *Date* ________ *Room* ________

Goals
1. Sketch the graph of a rational function.
2. Use rational functions as models.

State/Local Objectives ________________

NCTM Curriculum Standards: Problem Solving, Communication, Connections, Functions, Geometry

✓ **Check items you wish to use for this lesson.**

Introducing the Lesson
___ Teaching Tools: Problem of the Day copymasters page 49, or Teacher's Edition page 646
___ Teaching Tools: Warm-Up Exercises copymasters page 81, or Teacher's Edition page 646

Teaching the Lesson using the following:
___ Extension for Example 4, Teacher's Edition pages 647–648
___ Color Transparencies page 72
___ Teaching Tools: transparencies pages 4, 15

Notes for substitute teacher ________________

Closing the Lesson
___ Communicating about Algebra, Student's Edition page 648
___ Guided Practice Exercises, Student's Edition page 649

Homework Assignment, pages 649–651
___ Basic/Average: Ex. 7–11 odd, 13–18, 25–33 odd, 47–52
___ Above Average: Ex. 7–11 odd, 13–18, 23–33 odd, 39–40, 47–52
___ Advanced: Ex. 13–18, 19–35 odd, 37–40, 47–52

Reteaching the Lesson
___ Extra Practice Copymasters page 85
___ Reteaching Copymasters page 85

Extending the Lesson
___ Enrichment, Teacher's Edition page 651

Notes

Lesson Plan 12.6
pages 652–657

1 day for Basic Course, 1 day for Full Course

Teacher's Name ______________________ *Class* __________ *Date* ________ *Room* ________

Goals
1. Construct a stem-and-leaf plot.
2. Construct a box-and-whisker plot.

State/Local Objectives ____

NCTM Curriculum Standards: Problem Solving, Communication, Statistics, Discrete Mathematics

✓ **Check items you wish to use for this lesson.**

Introducing the Lesson
___ Teaching Tools: Problem of the Day copymasters page 49, or Teacher's Edition page 652
___ Teaching Tools: Warm-Up Exercises copymasters page 81, or Teacher's Edition page 652

Teaching the Lesson using the following:
___ Teaching Tools: transparencies page 4

Notes for substitute teacher ____

Closing the Lesson
___ Communicating about Algebra, Student's Edition page 654
___ Guided Practice Exercises, Student's Edition page 655

Homework Assignment, pages 655–657
___ Basic/Average: Ex. 5–13 odd, 19–26
___ Above Average: Ex. 5–19 odd, 20–27
___ Advanced: Ex. 5–19 odd, 21–27

Reteaching the Lesson
___ Extra Practice Copymasters page 86
___ Reteaching Copymasters page 86

Extending the Lesson
___ Enrichment, Teacher's Edition pages 657
___ Technology: Using Calculators and Computers page 87

Notes

Lesson Plan 12.7
pages 658–663

1 day for Basic Course, 1 day for Full Course

Teacher's Name ______ *Class* ______ *Date* ______ *Room* ______

Goals
1. Find the mean, median, and mode.
2. Use measures of central tendency in real-life situations.

State/Local Objectives ______

NCTM Curriculum Standards: Problem Solving, Communication, Connections, Statistics, Discrete Mathematics

✓ **Check items you wish to use for this lesson.**

Introducing the Lesson
___ Teaching Tools: Problem of the Day copymasters page 49, or Teacher's Edition page 658
___ Teaching Tools: Warm-Up Exercises copymasters page 82, or Teacher's Edition page 658

Teaching the Lesson using the following:
___ Color Transparencies pages 72, 73
___ Teaching Tools: transparencies pages 1, 2, 4

Notes for substitute teacher ______

Closing the Lesson
___ Communicating about Algebra, Student's Edition page 660
___ Guided Practice Exercises, Student's Edition page 661

Homework Assignment, pages 661–663
___ Basic/Average: Ex. 5–19 odd, 20–24, 25
___ Above Average: Ex. 5–11, 13–23 odd, 25
___ Advanced: Ex. 5–12, 17–26

Reteaching the Lesson
___ Extra Practice Copymasters page 87
___ Reteaching Copymasters page 87

Extending the Lesson
___ Enrichment, Teacher's Edition page 663
___ Technology: Using Calculators and Computers page 89

Notes

Lesson Plan 13.1
pages 676–681

1 day for Basic Course, 1 day for Full Course

Teacher's Name ______ *Class* ______ *Date* ______ *Room* ______

Goals
1. Find the distance between two points.
2. Find the midpoint between two points.

State/Local Objectives ______

NCTM Curriculum Standards: Problem Solving, Communication, Connections, Geometry

✓ **Check items you wish to use for this lesson.**

Introducing the Lesson
___ Teaching Tools: Problem of the Day copymasters page 50, or Teacher's Edition page 676
___ Teaching Tools: Warm-Up Exercises copymasters page 82, or Teacher's Edition page 676

Teaching the Lesson using the following:
___ Color Transparencies pages 75, 76
___ Teaching Tools: transparencies page 4

Notes for substitute teacher ______

Closing the Lesson
___ Communicating about Algebra, Student's Edition page 679
___ Guided Practice Exercises, Student's Edition page 680

Homework Assignment, pages 679–681
___ Basic/Average: Ex. 4–29 odd, 34–38, 39–49 odd
___ Above Average: Ex. 4–31 odd, 32–36, 39–49 odd, 50
___ Advanced: Ex. 9–31 odd, 32–36, 39–49 odd, 50

Reteaching the Lesson
___ Extra Practice Copymasters page 88
___ Reteaching Copymasters page 88
___ Alternative Assessment: Math Log pages 68–70

Extending the Lesson
___ Enrichment, Teacher's Edition page 681

Notes

Lesson Plan 13.2
pages 682–687

3 days for Basic Course, 3 days for Full Course

Teacher's Name ______________________ *Class* __________ *Date* __________ *Room* ________

Goals
1. Simplify radicals by applying their properties.
2. Use radicals in real-life situations.

State/Local Objectives ____

NCTM Curriculum Standards: Problem Solving, Communication, Structure, Connections, Geometry

✓ **Check items you wish to use for this lesson.**

Introducing the Lesson
___ Teaching Tools: Problem of the Day copymasters page 50, or Teacher's Edition page 682
___ Teaching Tools: Warm-Up Exercises copymasters page 83, or Teacher's Edition page 682

Teaching the Lesson using the following:
___ Color Transparencies page 76

Notes for substitute teacher ____

Closing the Lesson
___ Communicating about Algebra, Student's Edition page 684
___ Extend Communicating, Teacher's Edition page 685
___ Guided Practice Exercises, Student's Edition page 685

Homework Assignment, pages 685–687
___ Basic/Average: Ex. 7–35 odd, 41–42, 47–59 odd
___ Above Average: Ex. 9–37 odd, 39–46, 47–59 odd
___ Advanced: Ex. 11–39 odd, 40–47, 59–64

Reteaching the Lesson
___ Extra Practice Copymasters page 89
___ Reteaching Copymasters page 89

Extending the Lesson
___ Enrichment, Teacher's Edition page 687
___ Applications Handbook pages 49–51

Notes

Lesson Plan 13.3
pages 689–694

3 days for Basic Course, 3 days for Full Course

Teacher's Name ____________________ *Class* __________ *Date* __________ *Room* __________

Goals
1. Add and subtract radical expressions.
2. Use radical operations in real-life problems.

State/Local Objectives ______________________________

____

NCTM Curriculum Standards: Problem Solving, Communication, Connections, Geometry

✓ **Check items you wish to use for this lesson.**

Introducing the Lesson
___ Teaching Tools: Problem of the Day copymasters page 50, or Teacher's Edition page 689
___ Teaching Tools: Warm-Up Exercises copymasters page 83, or Teacher's Edition page 689

Teaching the Lesson using the following:
___ Cooperative Learning: Example 5, Teacher's Edition page 690
___ Color Transparencies page 77
___ Extension for Example 4, Teacher's Edition page 690
___ Teaching Tools: transparencies page 4

Notes for substitute teacher ______________________________

____

____

Closing the Lesson
___ Communicating about Algebra, Student's Edition page 691
___ Guided Practice Exercises, Student's Edition page 692

Homework Assignment, pages 692–694
___ Basic/Average: Ex. 7–29 odd, 33–34, 41–46
___ Above Average: Ex. 9–29 odd, 31, 33–36, 41–46
___ Advanced: Ex. 11–29 odd, 31–36, 41–46

Reteaching the Lesson
___ Extra Practice Copymasters page 90
___ Reteaching Copymasters page 90

Extending the Lesson
___ Enrichment, Teacher's Edition page 694

Notes

____

____

____

Lesson Plan 13.4
pages 695–700

1 day for Basic Course, 1 day for Full Course

Teacher's Name ______________________ *Class* __________ *Date* __________ *Room* __________

Goals 1. Solve radical equations.
2. Use radical equations to solve real-life problems.

State/Local Objectives ______________________________

NCTM Curriculum Standards: Problem Solving, Communication, Connections, Geometry

✓ **Check items you wish to use for this lesson.**

Introducing the Lesson
___ Teaching Tools: Problem of the Day copymasters page 51, or Teacher's Edition page 695
___ Teaching Tools: Warm-Up Exercises copymasters page 83, or Teacher's Edition page 695

Teaching the Lesson using the following:
___ Teaching Tools: transparencies pages 1, 2

Notes for substitute teacher ______________________________

Closing the Lesson
___ Communicating about Algebra, Student's Edition page 697
___ Guided Practice Exercises, Student's Edition page 698

Homework Assignment, pages 698–700
___ Basic/Average: Ex. 5–27 odd, 33–37, 45–49 odd
___ Above Average: Ex. 7–27 odd, 31–36, 39–49 odd
___ Advanced: Ex. 9–29 odd, 31–37, 45–51 odd

Reteaching the Lesson
___ Extra Practice Copymasters page 91
___ Reteaching Copymasters page 91

Extending the Lesson
___ Enrichment, Teacher's Edition page 699
___ Applications Handbook pages 1–5, 55, 56

Notes

Lesson Plan 13.5
pages 703–709

Omit for Basic Course, 1 day for Full Course

Teacher's Name ____________________ *Class* __________ *Date* __________ *Room* __________

Goals

1. Use the tangent of an angle to find the lengths of the sides of a right triangle.
2. Use the tangent to solve real-life problems.

State/Local Objectives ____________________

____

NCTM Curriculum Standards: Problem Solving, Communication, Connections, Geometry, Trigonometry

✓ **Check items you wish to use for this lesson.**

Introducing the Lesson
___ Teaching Tools: Problem of the Day copymasters page 51, or Teacher's Edition page 703
___ Teaching Tools: Warm-Up Exercises copymasters page 84, or Teacher's Edition page 703

Teaching the Lesson using the following:
___ Teaching Tools: transparencies pages 11, 12

Notes for substitute teacher ____________________

____

____

Closing the Lesson
___ Communicating about Algebra, Student's Edition page 705
___ Extend Communicating: Research, Teacher's Edition page 706
___ Guided Practice Exercises, Student's Edition page 706

Homework Assignment, pages 706–709
___ Basic/Average: Ex. 5–15 odd, 19–27 odd, 33–34, 37–43
___ Above Average: Ex. 5–27 odd, 28, 31–34, 38–45
___ Advanced: Ex. 5–27 odd, 28–35, 41–45

Reteaching the Lesson
___ Extra Practice Copymasters page 92
___ Reteaching Copymasters page 92

Extending the Lesson
___ Enrichment, Teacher's Edition pages 708–709
___ Cultural Diversity Extensions page 13

Notes

____

____

____

Lesson Plan 13.6
pages 710–715

Omit for Basic Course, 1 day for Full Course

Teacher's Name ______________________ *Class* __________ *Date* __________ *Room* __________

Goal Use properties of algebra to prove a statement true and use counterexamples to prove a statement false.

State/Local Objectives ____

NCTM Curriculum Standards: Problem Solving, Communication, Connections, Structure, Reasoning

✓ **Check items you wish to use for this lesson.**

Introducing the Lesson
___ Teaching Tools: Problem of the Day copymasters page 51, or Teacher's Edition page 710
___ Teaching Tools: Warm-Up Exercises copymasters page 84, or Teacher's Edition page 710

Teaching the Lesson using the following:
___ Color Transparencies page 78

Notes for substitute teacher ____

Closing the Lesson
___ Communicating about Algebra, Student's Edition page 712
___ Extend Communicating: Research, Teacher's Edition pages 712–713
___ Guided Practice Exercises, Student's Edition page 713

Homework Assignment, pages 713–715
___ Basic/Average: Ex. 5–9 odd, 11–12, 15–21 odd
___ Above Average: Ex. 5–11 odd, 12–14, 15–23 odd
___ Advanced: Ex. 5–11 odd, 12–14, 15–23 odd

Reteaching the Lesson
___ Extra Practice Copymasters page 93
___ Reteaching Copymasters page 93

Extending the Lesson
___ Enrichment, Teacher's Edition page 715

Notes